AF559656

Protected Nutrient Technology for Crossbred Cattle

About the Authors

Dr. S.H. Mane, He did Doctor of philosophy in Animal Husbandry and Dairy Science from Mahatma Phule Krishi Vidyapeeth, Rahuri (Maharashtra). He has worked as Senior Research Assistant at College of Agriculture, Dhule and Subject Matter Specialty (Animal Science) at Krishi Vigyan Kendra, Dhule. He has authored two books on Fodder Production and Milk and Milk Products processing for farmers, 18 Research papers, Abstract in National and International Journals along with 150 popular articles. He got "Krishi Bhushan" award by Shriyash Foundation Pune. He is actively engaged in Teaching, Research and Extension activities from last 10 years. At present, he is working as Assistant Professor of Animal Husbandry and Dairy Sciences, College of Agriculture, Pune (Maharashtra State)

Dr. Y. G. Fulpagare, M.Sc. (Agriculture), Ph.D. (Animal Nutrition), CCS Haryana Agriculture University, Hissar. He has worked as Head, Department of Animal Husbandry and Dairy Science, Mahatma Phule Krishi Vidyapeeth, Rahuri, Maharashtra. He had a experience of 34 years. He guided 33 students which includes 27 M.Sc. (Agril.) and 7 Ph.D. Students. He published 31 Scientific, 59 Technical and 18 Extension publications. Prof. Fulpagare has been a teacher, researcher, field worker, author-editor and consultant. He carried out several assignments on Animal Nutrition and Livestock production aspects. Presently he is Associate Dean, College of Agriculture, Dhule (Maharashtra).

Protected Nutrient Technology for Crossbred Cattle

S.H. Mane
Y.G. Fulpagare

NEW INDIA PUBLISHING AGENCY
New Delhi – 110 034

NEW INDIA PUBLISHING AGENCY
101, Vikas Surya Plaza, CU Block, LSC Market
Pitam Pura, New Delhi 110 034, India
Phone: + 91 (11)27 34 17 17 Fax: + 91(11) 27 34 16 16
Email: info@nipabooks.com
Web: www.nipabooks.com

Feedback at feedbacks@nipabooks.com

ISBN: 978-93-85516-60-3

National Research Centre on Camel
(Indian Council of Agricultural Research)
Post Bag-07, Jorbeer, Bikaner 334001 (Raj.) India

Dr. N.V. Patil
M.V. Sc., Ph.D.
Director

Foreword

The importance of livestock sector alone in contributing nearly 25.6% of Value of Output at current prices of total value of output in Agriculture, Fishing & Forestry sector is well known. The overall contribution of Livestock Sector in total GDP during 2012-13 was nearly 4.11% at current prices. During the last 5 decades, our nation's milk producers have transformed Indian dairying from stagnation to world leadership due to remarkable growth in milk production to nearly 145 Million MT has not only improved economy but also provided gainful employment to rural folks. However quite often Indian dairy stock is blamed for lower productivity and an increase in milk production is considered due to more livestock population. But as per the XIX th Livestock Census, although the overall total Bovine of 299.9 million population in 2012 has registered a decline in numbers—a decline of 1.57% over previous census. The number of milch animals (in-milk and dry) in cows and buffaloes has increased from 111.09 million to 118.59 million, an increase of 6.75% and increased number of animals in milk in cows and buffaloes from 77.04 million to 80.52 million showing a growth of 4.51% has indicated the trend adopted by farmers of keeping better animals for milk purpose.

The feeding management plays an important role in sustaining higher milk production in high producing dairy animals and especially nutrition during transition period of dairy animals during 3 weeks prior and 3 weeks after parturition is critically important to health, production and profitability of dairy. Nutrition and management of cows during the transition period has received tremendous interest in recent years in developed countries but relatively less attention is being paid in the developing countries like India. With understanding of role of energy and protein nutrition in managing the transition period some technologies like protected fat and protein alone or in combination has been practiced by few dairy farmers but many are still ignorant about it. The experimental results obtained with the use of these technologies have been discussed in greater depth in this book with emphasis on its profitable effect on early lactation performance and post partum reproduction. The role of energy nutrients as by pass fat and protein as by pass protein in isolation and in combination has also been discussed in the light of regaining the positive energy balance state after the parturition and its association with increased milk yield and improved post partum reproduction.

It is hoped that in changing situation of keeping higher number of milch animals by farmers it will be appropriate for them to understand the importance of feeding dairy animals during transition phase and exploit the milk production function optimally along with proper care and attention towards the health of dairy animals so that it continues to impart sustenance to profitability to dairy enterprise.

(N.V. Patil)

Phone : +91-151-2230183 (O), Fax: +91-151-2231213, Mobile:09414721588
e-mail: nvpatil61@gmail.com,nrccamel@nic.in website : www.nrccamel.res .in

Acknowledgements

I feel extremely honoured for the opportunity bestowed upon me to work under the versatile guidance of my honourable research guide Dr. Y. G. Fulpagare, Senior Scientist, RCDP on cattle and I/C Head, Department of Animal Husbandry and Dairy Science, Mahatma Phule Krishi Vidyapeeth, Rahuri. Indeed, my indebtedness to him becomes unquantifiable for his personal interest and scholastic guidance, constant encouragement and constructive criticism throughout the course of this investigation and presentation of this dissertation.

I am highly obliged to Dr. B. R. Ulmek, Director of Instructions and Dean, Faculty of Agriculture, M. P. K. V., Rahuri for providing necessary facilities and supports during the course of my study in the University.

I would also like to place on record my sincere thanks and deep sense of gratitude to the members of Advisory Committee, Dr. S. D. Mandakmale, Senior Scientist, AICRP on Goat (Sangamneri) field unit, Department of Animal Husbandry and Dairy Science, Dr. D. K. Deokar, Assistant Dairy Cattle Breeder, RCDP on cattle, M.P.K.V., Rahuri and Dr. C.A. Nimbalkar, Associate Professor, Department of Agricultural Economic and Statistics, M.P.K.V., Rahuri for their valuable guidance and help during my course of investigation.

I am grateful to Prof. A. B. Deshmukh, Librarian, College of Agriculture, Pune for making library facilities available for the present investigation.

I am deeply obliged to all other Scientist past and present whose literature has been cited sacrifice, constant encouragement and conciliation have boosted up me towards cherished dreams are turning into reality in the form of dissertation.

I cannot find word to express, heartiest gratitude to my family, father Appa, mother Aai, Wife Mrs. Asha, my sweet daughters Miss. Sonam and Kranti, my Brothers Gajanan, Sachin, Manoj and Ravindra and nephews Samarth, Soham, Ram, Krishna, niece Harshada and grandmother Smt. Parvathibai for the unceasing encouragement, support and attention. These affectionate people have always inspired me to achieve success and above all they have given me moral strength to make myself a successful man. Also, I wholeheartedly express my deep sense of love and affection to my wife Late. Revati for her loving contribution in my personal as well as my academic life.

I am deeply indebted to Late. Dr. Padmashri, Appasaheb Pawar who supported and encouraged me to take up education in the field of Agriculture and serve the farming community. I dedicate all that I am to him and to the First Family of Baramati: Late. Dr. Padmashri, Appasaheb Pawar and the members of the family.

I owe my deepest gratitude to Honble Ajit Dada Pawar, Sau. Sunetra Vahini Pawar and Rajendra Dada Pawar, Sunanda Vahini who always encourage and supported my educational and academic progress.

Place: Pune S.H. Mane

Contents

List of Tables

List of Colour Plates

Chapter 3: Material and Methods

List of Figures

Abbreviations

%	:	Per cent
@	:	At the rate of
°C	:	Degree Celsius
C. D.	:	Critical difference
cm/d	:	Centimeter/ day
Con. Mix.	:	Concentrate Mixture
DCF	:	Digestible Crude Fiber
DCP	:	Digestible Crude Protein
DEE	:	Digestible Ether Extract
DMI	:	Dry matter intake
DNFE	:	Digestible Nitrogen Free Extract
et al.	:	And other (et alia)
etc.	:	Et cetera (and so on / and rest of)
FA	:	Formaldehyde
FCE	:	Feed Conversion Efficiency
Fig.	:	Figures
g/d	:	Gram/day
GNC	:	Groundnut cake
HCHO	:	Formaldehyde
hr.	:	Hours
kg/d	:	Kilogram/day
N	:	Nitrogen
N.S.	:	Non-significant
qt	:	quintal
RDP	:	Rumen degradable protein
Rs.	:	Rupees
S.E.	:	Standard error
TA	:	Total ash
TDN	:	Total Digestible Nutrients
UDP	:	Undegradable protein
viz.,	:	Namely

Abstract

The present experiment entitled, "Protected Nutrient Technology for Crossbred Cattle" was conducted at Research Cum Development Project on Cattle, Department of Animal Husbandry and Dairy Science, Post Graduate Institute, Mahatma Phule Krishi Vidyapeeth, Rahuri Dist. Ahmednagar, Maharashtra

Major objective was to study the nutrient utilization, production and reproduction performance of crossbred cattle by feeding protected fat and protein. The feeding trial of 4 months duration (1 month prepartum and 3 month postpartum) was conducted on 24 crossbred cattle which were subjected to 4 treatments and designated as treatment T_0, T_1, T_2 and T_3. Crossbred cows in second to fourth lactation with most probable production ability (MPPA) of average approximately 2300 liter milk production per lactation for each group were selected. All animals were provided/fed with 2/3 DM through roughages (2/3 from dry roughages and 1/3 from green roughages) + 1/3 DM from concentrate mixture. In T_0 and T_2 groups untreated groundnut cake was given in ration. Whereas, in T_1 and T_3 groups, untreated groundnut cake was replaced with formaldehyde (FA) (@ 1.0 gm FA /100g CP) treated groundnut cake. Also bypass fat (99%) was supplemented in T_2 and T_3 groups @ 10 gm per liter milk production. The data generated during experimental period were analyzed by Factorial Randomized Block Design (FRBD)/ SAS, 9.3 versions.

The experimental ration contained 18.61, 18.65, 18.40 and 18.55 per cent crude protein in T_0, T_1, T_2 and T_3, respectively which indicated that, all rations prepared for different treatments were almost isonitrogenous.

After go days of the calving of experimental cows, there was net weight loss of 13.3 kg in T_0 while, gain in weight was seen in T_3 (22 kg) followed by T_2 (15) and T_1 (4.17 kg), respectively. It indicated that weight gain in groups T_3 and T_2

"Protected Nutrient Technology for Crossbred Cattle" by
SOMNATH HANUMANT MANE
M.Sc (Agri.) Ph.D.
In ANIMAL HUSBANDRY
Research Guide : Dr. Y.G. Fulpagare
Department : Animal Husbandry and Dairy Science

attained positive energy balance earlier as was evident from increasing trend in body weights while it took longer (3rd fortnight) time in case of group T_1 and (4th fortnight) in group T_0. The BCS was also significantly higher in T_3 (3.64) followed by T_2 (3.60), T_1 (3.57) and T_0 (3.48), respectively. However the average birth weights of the calves born of experimental dams were non significant in all groups.

The DMI was significant, ($P<0.05$) higher in T_3 (12.72) followed by T_1 (12.59), T_0 (12.47) and T_2 (12.23). However the DMI/ 100 kg body weight was 3.05, 3.01, 3.13 and 2.81 kg/d in T_0, T_1, T_2 and T_3, respectively which was non significant among all treatment. There was no effect on the digestibility of DM and CP. However, CF and EE digestibility was higher ($P<0.05$) fed with in all treatment group than control. NFE digestibility was highest in group protected fat and protein combination (T_3) and was followed by group fed protected fat alone (T_2). Treatment T_3 and T_2 were at par and treatment T_1 had significantly lower NFE digestibility.

The average digestible crude protein intake (DCPI) and total digestible nutrients intake (TDN) was significantly higher ($P<0.01$) in T_3, T_2 and T_1 than T_0 due to higher DMI. However DCPI and TDNI / 100 kg body weight did not differ among all groups.

Average total digestible nutrients intake was (TDN) significantly higher ($P<0.01$) in T_3 followed by T_2 and T_1 over T_0. The TDNI/ 100 kg body weight were 1.74, 1.88, 2.03 and 1.84 kg/day in T_0, T_1, T_2 and T_3 treatment respectively showing the similar trend as that of DCP and DCPI/ 100kg body weight.

The average milk production during supplementation period was 9.82, 11.76, 11.41 and 12.43 kg/d in group T_0, T_1, T_2 and T_3 respectively which was significantly 16.49% higher in T_1, 13.93% higher in T_2 and 20.99% in T_3 over T_0. Similar trend was observed in 4% FCM and ECM yield in all treatment groups.

There was significant difference ($P<0.05$) in milk fat, milk protein, lactose, total solids, SNF , density, temperature and mineral matter among all treatment groups by feeding protected protein and fat. The supplementation of protected protein and fat increased the proportion of unsaturated and long chain fatty acids of milk while saturated fatty acids content was decreased.

There was significantly less ($P<0.05$) time required for expulsion of fetal membrane in T_2 (2.3), T_3 (1.88) and T_1 (1.7) over T_0. Similarly, there were ($P<0.05$) significantly less days required for involution of uterus in T_3 (5.83 days) and T_2 (3.16 days) than T_0 and T_1 (0.83). The time required for commencement of first heat after calving was reduced by 26 days in T_3, 21.83 days in T_2 and 11.83 days in T_1 over T_0. There was statistically significant

($P<0.01$) difference in T_3 and T_2 over T_0 but T_1 and T_0 were at par. The average service period was significantly lower in T_2 (82.00) than all treatment groups, followed by T_3 (99.67), T_1 (130.67) and T_0 (141.33) days. The average numbers of AI required were 3.00, 2.50, 1.83 and 2.00 in T_0, T_1, T_2 and T_3, respectively. Highest conception rate per cent was found in T_2 (66.67) treatment group followed by T_3 (54.55), T_1 (40.00) and lowest in T_0 (33.34). Four cases of retention of fetal membranes were observed in T_0 and 3 cases in T_1 and one case each in T_2 and T_3 group. Similarly, 2 cases of metritis were observed in T_0 and one case in T_1.

There was no effect on albumin, globulin, total blood protein, NEFA, uric acid and creatinine among treatments. The average BUN level was significantly lower ($P<0.01$) in T_1 and T_3 compared to T_0 and T_2 treatments groups. The cholesterol level was higher in group T_3 and T_2 as compare to T_0 and T_1 which was statistically significant. The cholesterol level was significantly lower in control group (T_0) over all treatment groups. The blood glucose level remained significantly higher in treatment groups over control . The highest ($P<0.05$) mean value of triglycerides was observed in T_3 followed by T_2, T_1 and T_0 mg/dl respectively.

The results of the present study indicated that feeding of different protected protein, fat or combination of both significantly improved digestibility of EE, CF and NFE. Daily milk yield was increased by 16.49, 13.93 and 20.99 per cent by feeding protected protein; protected fat and combination of both respectively over control. Milk composition was affected by different feeding treatment. Similarly, total yield was also significantly higher in treatment group. Reproductive performances of animal were positively improved. By supplementation of protected fat, blood cholesterol and triglycerides level are significantly increased and BUN level decreased significantly by feeding protected protein. Protected protein and fat combined treatment showed significantly better production and reproduction performance than other treatments. Hence the feeding of protected protein and fat could be advised for better monitory returns from milking cows.

Chapter 1

Introduction

The livestock sector is one of the fastest growing agriculture sub sector in India, which accounted for about 17% of the value of agriculture output in 1981-83, has gradually increased by to 25% in 2011-2012 (CSO 2011, GOI). India possesses one of the largest livestock populations in the world with 218 million cattle, 115.4 million buffalos, 160 million goat, 75 million sheep, 9.4 million pigs, 947 million poultry (FAO, 2012) which provide excellent opportunities and boost rural income and alleviation of poverty. The country has 16.5% cattle, 56.8% buffalo, 5.78% sheep and 14.76% goat population of the world. India stands first in milk production likely to be touch142.6 MT in 2014 with 4.5% annual growth rate of milk. The per capita availability of milk rose from 112 g/head/day in 1970 to 296 g/head/day 2013-14 (NDDB. 2014). Owing to this huge bovine stock, though India has managed to attain *numero uno* position in milk production, the full potential of Indian milch herd remains unattained. Over the last three decades (1982 to 2012), average productivity of Indian cattle and buffaloes has grown from 1.9 to 3.9 kg per day and from 3.7 to 6.2 kg per day, respectively. The average daily milk yield for crossbred cattle is better at 7.1 kg per day, but still significantly lesser than the best of global standards viz. UK, US and Israel are at 25.6, 32.8 and 38.6 kg per day, respectively. The major causes of low productivity in India are both intrinsic (low genetic potential) and extrinsic such as poor nutrition/feed management, inferior farm management practices, ineffective veterinary and extension services and inefficient implementation of breed improvement programmes.

In order to meet growing milk demand of 170 MT in 2030, India's milk output has to grow by 6 millions tonns per annum(MTPA) every year for next 10 years, failing to achieve this world's largest milk producer may have to resort to import. (Hon'ble Dr. Radha Mohan Singh, Union Agri. Min. GOI, Ref. The Hindustan Business Line, 3.7.2014).

India must increase the average incremental milk production from 4 million (mt.) in the last 10 years to 7.8 mt in the next 8 years to meet the growing demand. India's estimated demand for milk is expected to be about 155 mt by 2016-17 and 200 mt in 2021-22 (National Dairy Development Board, 2014).

Adequate nutrition is most important for maintenance, growth and reproduction of animals. Genetic improvement accounts for 33-40% of this increase while feeding and management contributes the remaining 60-67%. Indian livestock are underfed and undernourished. At present, the country faces a net deficit of 61.1% green fodder, 21.9% dry crop residues and 64% feeds. The deficit and supply in crude protein (CP) and total digestible nutrients (TDN) are 28 and 24%, respectively (10th five year plan). This is due to the perhaps huge number of animals in relation to available feed and fodder resources. (Jadhav *et al.*, (2014) Animals largely derive their feed requirements from crop by-products and grazing on common lands. Further, a majority of livestock is owned by small landholders and the landless and often face acute shortage of feed and fodder. Animal science research has yielded a number of biochemical and mechanical technologies that besides improving the nutritional quality of feed also help to avoid wastage of these resources. The efficacy of many of these techniques has been proved, yet their adoption has remained restricted and sporadic due to a number of operating constraints.

The productivity and profitability in any livestock sector to a large extent is determined by feed resources and the quality of the feeds available as feed is the single largest recurring expenditure accounting for 60-75% of the cost of production. Rapidly growing livestock sector necessitates a proportional increase in the availability of feed resources to fulfill its demand. Limited availability of natural resources i.e. land and water and the growing demand for food and commercial crops are some of the major constraints in expanding the feed resources. Under the present circumstances of a limited feed resources on the supply side and greater demand for livestock products on the demand side, requiring for higher quantity and quality of feed resources and therefore use of unconventional feed resources in animal feeding is now days becoming important.

India, with only 2.29 % of land area of the world, is maintaining nearly 17.4% of human and 10.7% of livestock population of the world. Growing demand for livestock products by the increasing population and urbanization is one of the major factors responsible for the steady growth of the livestock sector. Sustaining the current growth trends in the livestock sector and fulfilling the demand for feed without compromising on food security and any additional allocation of land and water for feed production would be a major challenge for the livestock sector. In view of the current shortage and the likely chances of further escalating deficits of feed resources there is a need for a concrete action plan for the feed resource management for the livestock sector. Resource wise dry fodder represents the largest amount followed by the green fodder and the concentrates. Dry fodder mostly comprises of crop residues and is obtained as a by-product after harvesting the food grains. Crop residues have low palatability, nutritionally poor and have to be supplemented with green fodder or concentrates to maintain productive animals. Green fodders are the next largest feed resource and are highly palatable, nutritionally rich and can be fed to sustain medium to reasonably good level of production. Concentrates comprise of grain by-products (bran and chunni), oil cakes and grains and are the most desired class of feed resources for sustaining high levels of production.

Most of the animals in developing countries including India are fed on agriculture by-products and low quality crop residues, which have got inherent low nutritive value and digestibility. The shortage of feed resources coupled with their poor nutritive value is of major concern to low productivity of dairy animals. Livestock production contributes significantly to rural economy and could be cash crops in many small holder mixed farming systems. In tropical countries, there is horizontal growth in terms of animal numbers and now there is need to achieve vertical growth in terms of improving productivity, so that future demand of milk would be met. This can be achieved, if the early lactating high yielding and genetically improved cows and buffaloes are fed according to the nutrient requirement with high energy diet. In tropical countries, the majority of livestock subsist on poor quality native grasses, crop residues and agro-industrial byproducts. Therefore, high yielding and genetically improved dairy animals pose a big challenge to provide the essential nutrients for meeting metabolic requirements and sustaining milk production.

Milk yield and optimum reproductive performance are the most important factors in determining profitability of dairy animals and high milk production is always more important for high profitability than the low feeding cost. In early lactating cows the energy intake through ration doesn't meet the requirement for higher milk production, resulting in a Negative Energy Balance (NEB), which is closely related to reproductive performance (Shelke *et al.*, 2011). Therefore, minimizing the extent and duration of NEB could be beneficial for reproduction besides getting the best productive performance from cows (Tyagi *et al.*, 2010). Nutrient requirements of dairy cows vary tremendously with stage of lactation and stage of gestation, therefore, nutrient density must change and feed composition needs to be readjusted periodically during the lactation cycle. Deleterious effect of NEB on productive performance of early lactating animals would be reduced by supplementation of protected fat in the ration through enhancing energy intake (Remppis *et al.*, 2011). Protected protein feeding to lactating animals leads to proportionate increase in the supply of amino acids to the host ruminant for productive/ reproductive purpose, with an overall increase in the efficiency of protein and energy utilization.

Protected nutrient technology is one such approach, involving feed management through passive rumen manipulation, by which the dietary nutrients (fat and protein) are protected from hydrolysis, allowing these nutrients to bypass rumen and get digested and then absorbed from the lower tract. The protected nutrients mainly include protected fat and protein and it is also called as bypass nutrients. Bypass nutrient mean that the essential and more important nutrient (high BV) should be escape from the rumen of face minimum fermentation. In India, there is a shortage of both energy as well as proteinaceous feeds. Protein supplements are more expensive and increase the feed cost. By optimizing the use of protein supplement within the ruminant system, we can reduce the quantity of protein in the diet and also can enhance the production of the animals. The protein can be divided in two parts, for the ruminant animals, in most of the feed, major part is degradable in rumen 'Rumen Degradable Protein' (RDP) and a small but variable amount of dietary protein escape rumen degradation 'Un-degradable Dietary Protein' (UDP). UDP which enters the lower tract is absorbed mostly as amino acids following enzymatic digestion. Of the RDP fraction, substantial part is utilized as the nitrogen source for rumen microbes for protein synthesis, while the rest is absorbed as ammonia. Only part of absorbed ammonia is recycled back to rumen as urea via saliva, the rest excreted out through urine. The host animal gets amino acids requirement from two sources i.e. microbial protein and undegradable protein (UDP), both flowing to lower tract. Although in the case of low yielder, the microbial protein synthesized in the rumen is sufficient. In some cases microbial protein and UDP is sufficient to meet animal's requirement. In fast growing animal and high yielding animals microbial supply is limited than the demand of amino acids at the tissue level, so to support the demand, it is necessary to provide proteins in the form of UDP or escape proteins or protected proteins There is limited availability of protein supplements like cakes/meals in India. Total protein available from concentrate in India is estimated to be around 8.5-9 million tones which can support production of only 0.45 million tons of milk proteins by the present mode of its utilization. Studies showed that when Protein degradation in rumen is controlled and it is made to bypass, the same can support the production 1.72 millions of milk proteins. However, if these resources are given suitable treatments, their degradation in the rumen may be controlled and thus, their efficiency of utilization for growth and milk production can be substantially increased (Garg *et al.*, 2003a, Walli, 2005). Highly degradable proteinous oil cakes when ingested by ruminants, result in large scale ammonia production, much of it gets wasted as urea excreted through urine. Even the animal has to spent energy to convert ammonia into urea in liver. In order to increase the efficiency of protein utilization from the highly degradable cakes, these proteins need to be protected from excessive ruminal degradation and can be used as protected protein, so that the amino acids from these protein feeds are absorbed intact from the intestines of the animal for tissue protein synthesis as well as for the process of gluconeogenesis in liver (Walli, 2005).

The transition period is very critical period in dairy cows is deûned as the last 3 week prepartum until 3 week postpartum and is characterized by marked changes in metabolism to support late gestation and the onset of milk synthesis. Along with a gradual decline in dry matter intake (DMI) that starts 2–3week prepartum, an abrupt increase in nutrient demand with initiation of lactation results in negative energy balance (NEBAL) and extensive mobilization of body fat reserves as non-esteriûed fatty acids (NEFA). Depressed concentrations of glucose and frequently insulin are typical during NEBAL in association with increased hepatic gluconeogenesis. NEBAL has detrimental effects on reproduction and production. Cows that lose body condition score (BCS) extensively during the ûrst 30 days of lactation experience long intervals to ûrst ovulation and concentration of circulating Beta Hydroxy Butarate and NEFA were greater in cows in which ovulation did not occur from the ûrst postpartum dominant follicle that developed after parturition. In high producing dairy cows, especially during early lactation, the amount of energy and protein required for maintenance of body tissues and milk production often exceeds the amount of energy available from diet (Goff and Horst, 1997). This condition arises because of reduced feed intake due to the stress of calving and milk production. The maintenance of lactation of dairy cows depends on the balanced feeding during postpartum period. It has been found that hydrogenated palm oil triglyceride provide a better energy supply for high-yielding dairy cows in negative energy balance than calcium soaps of palm oil fatty acids around calving (Karcagi *et al.*, 2010) and (Singh *et al.,* 2014).

Researchers have attempted to obtain milk fat with healthier properties increasing its content in polyunsaturated fatty acids (PUFAs) which have unquestionable beneficial effects on human lipid metabolism. Many attempts have been made, therefore, to alter the fatty acid composition of milk fat from lactating cows in order to improve its nutritional value.

The utilization of dietary proteins must be put in the context of the available energy supply. Energy is the main driving force of metabolism. If energy is limiting dietary protein will be used inefficiently as another source of energy instead of being converted into body protein. Protein synthesis in the body involves a considerable expenditure of energy to create the activated amino acids to be linked together. In addition, protein tissues are constantly being turned over.

When additional energy is provided, there is an increase in protein synthesis and a decrease in protein degradation and these two effects combine to enhance net protein retention. When additional protein is supplied at constant energy, there is an increase in both protein synthesis and in protein degradation, resulting in a smaller net increment in protein retention.

In developing countries like India, supplementation of protected fat and protein is beneficial to medium and high yielding cows and buffaloes but the cost effectiveness of the same needs to be kept in mind. As about the feeding of protected protein and fat, the results of some farm and field studies have indicated the usefulness and cost effectiveness of its feeding to cows and buffaloes yielding around 5-8 L of milk. In addition, milk fat yield and percentage of unsaturated fatty acids in milk fat was increased, resulting in improvement in nutritive value of milk, from a human health point of view. The supplementation of energy along with undegradable protein in the form of formaldehyde treated cake is beneficial in terms of increase in live weight gain, production and reproduction performance of animal. The importance of feeding bypass protein alone or in combination with bypass fat that results in improvement in dry matter intake, yield of whole milk, 6% FCM, fat and TS and return over feed cost from buffaloes during their early lactation (Vahora *et al.*, 2013).

With this view, the present study was conducted to investigate the **"effect of supplementing protected fat and protein alone and in combination on productive and reproductive performance in crossbred cattle"** with the following objectives.

i. To evaluate the nutrient utilization in crossbred cattle fed on protected fat and protein.

ii. To determine the effect of protected fat and protein on milk production and its composition.

iii. To assess the effect of protected fat and protein on reproduction performance.

iv. To study the effect of protected fat and protein on blood constituents profile.

Chapter 2

Review of Literature

In recent years, there has been a considerable interest in the new technological interventions for enhanced livestock productivity in the Indian sub-continent, because of the increased demand for livestock products for human consumption. Bypass nutrients defined as a nutrients fed in such a form that provides an increase in the flow of the nutrient (s) unchanged to the abomasum, yet is available to the more important nutrient (high BV) should escape from the rumen or face minimum ruminal fermentation. The reasons for protecting different nutrients are based on their mode of utilization in the rumen. The literature on the effect of the bypass fat and protein on nutrient utilization, milk production and reproductive performance has been reviewed and presented in following subheads

2.1 Role of the Protected Fat

Role of the protected fat in the rations of the high producing dairy animals is very crucial for enhancing the energy density of ration (NRC, 2001). Dietary fat, that resists lipolysis and biohydrogenation in rumen by rumen microorganisms, but gets digested in lower digestive tract, is known as rumen protected fat or bypass fat or inert fat.

2.2 Methods of Fat Protection

2.2.1 Natural bypass fat

Whole oil seeds, when fed without processing except drying have natural bypass fat properties due to their hard outer seed coat, which protects the internal fatty acids from lipolysis and biohydrogenation in rumen (Ekeren *et al.*, 1992). However, during mastication by animals there is physical breakdown of seed coat, which gives poor result of rumen inertness. Important whole oil seeds commonly used in the ration of dairy animals are cotton, roasted soybeans, sunflower and canola. Further, feed ingredients containing saturated fatty acids are less toxic to the ruminal microorganisms and minimize the adverse effects of the fat supplementation as they react more readily with the metal ions forming insoluble salts in rumen (Jenkins and Palmquist, 1982) and do not go for further ruminal biohydrogenation (Chalupa *et al.*, 1986).

2.2.2 Chemically prepared bypass fat

Chemically prepared bypass fat mainly includes crystalline or prilled fatty acids, formaldehyde treated protein encapsulated fatty acids, fatty acyl amides and calcium salts of long chain fatty acids (Ca-LCFA) and hydrogenate fat.

2.2.3 Crystalline or prilled fatty acids

Crystalline or prilled fatty acids can be made by liquifying and spraying the saturated fatty acids under pressure into cooled atmosphere, so that melting point of the fatty acids is increased and do not melt at ruminal temperature, thus resisting rumen hydrolysis and association with bacterial cells or feed particles (Chalupa *et al.*, 1986).

2.2.4 Formaldehyde Treated Protein encapsulated fatty acids

Formaldehyde treated protein encapsulated fatty acids is also an affecting means of protecting dietary fat from rumen hydrolysis (Sutton *et al.*, 1983). Oil seeds can be crushed and treated with formaldehyde (1.2 g per 100g protein) in plastic bags or silos and kept for about a week.

2.2.5 Fatty acyl amide

Fatty acyl amide can be prepared and used as a source of bypass fat. Butylsoyamide is a fatty acyl amide consisting of an amide bond between soy fatty acids and a butylamine, which increases linoleic acid content of the milk fat (Jenkins, 1998). Conversion of oleic acid to fatty acyl amide (oleamide) increases the mono-unsaturated fatty acids concentration of the milk, when fed to dairy cows (Jenkins, 1999). Fatty acyl amide of sardine oil based complete diet is effective in protecting fat from degradation in rumen and improves the apparent and true dry matter degradability (Ambasankar and Balakrishnan, 2011).

2.2.6 Calcium salts of long chain fatty acids

Calcium salts of long chain fatty acids (Ca-LCFA) are insoluble soaps produced by reaction of carboxyl group of long chain fatty acids (LCFA) and calcium salts (Ca^{++}). Degree of insolubility of the Ca soaps depends upon the rumen pH and type of fatty acids. When rumen pH is more than 5.5, Ca-LCFA is inert in rumen. As dissociation constant (pKa) of Ca-LCFA is 4 to 5, dissociation is significant, when pH decreases to 6.0 (Chalupa *et al.*, 1986). In acidic pH of the abomasum, fatty acids are dissociated from Ca-LCFA and then absorbed efficiently from small intestine. The unsaturated soaps are less satisfactory for maintaining normal rumen function, because dissociation is relatively higher. (Sukhija and Palmquist, 1990). Among all forms of bypass fat, Ca-LCFA is relatively less degradable in rumen has highest intestinal digestibility (Dairy Technical Service Staff, 2002) and serve as an additional source of calcium (Naik *et al.*, 2007a; 2007b).

2.2.7 Fractionated fat

Partially hydrogenated fat : 1st generation, hydrogenation of animal & vegetable oils (1980)

CaSFA : 2nd generation, calcium salt of fatty acids (1990)

Stable fats : 3rd generation, hydrolyzed and fractionated fats (Pure free fatty acids)

Functional fats : 4th generation, vitamins-enriched stable fats rumen bypass or 'protected' fats are dry fats that are processed to be easily mixed into animal feeds. Because dry fats naturally have high melting points, they are mostly insoluble at rumen temperature. In essence, dry fats are not as much 'protected' as completely insoluble in the rumen such that they have little impact on rumen fermentation. Today, there are only three methods of producing dry fats for animal feeds. The method that produces the least desirable product for the cow, partial hydrogenation of tallow, is seldom used for dairy rations. One acceptable method for producing a bypass fat is to hydrolyse the fatty acids from tallow, partially hydrogenate them, and then prill them in a spray-chilling tower. The most widely used and effective method for producing a rumen bypass fat is to react vegetable fatty acids with calcium oxide to form insoluble calcium soaps.

Three main types of rumen inert fats currently used in lactating dairy cow diets are: partially hydrogenated tallows (PHT), Ca salts of fatty acids (CaSFA), and hydrogenated free fatty acids (FFA). These fat types were developed to be used in dry form to provide dairy producers with a more functional physical product and to facilitate on- farm handling. Many commercially available rumen protected fats sold in the market. They contain a fat between 80-99%. These fats are specifically processed products that provide fat as their prime nutrient. These fats are commonly referred to as ruminal inert fat, protected fat, escape fat, and bypass fat and are more expensive per unit of energy provided compared to commodity fats. Commodity fats can affect rumen fermentation by absorbing bacteria and feed particles coating the feed or lower feed digestibility. Unsaturated fatty acids are more toxic because they bind more to the bacteria and impact the rumen fermentation. Exposure of the unsaturated fatty acids in the rumen due to oilseed processing (whole seed, rolled, ground, or extruded) or oil will impact field results.

2.2.7.1 First generation fats

Partially hydrogenated tallows are the first generation of rumen inert fats. They are produced by hydrogenating tallow or vegetable fats to increase the melting point of the end product. Tallow or vegetable fats may contain as much as 85% unsaturated fatty acids prior to biohydrogenation and as little as 15% after the hydrogenation process. The iodine value, an indicator of the degree of unsaturation, can vary from 14 to 31. Hydrogenation of tallow and vegetable fats reduces negative effects that fatty acids have on rumen fermentation. However, the same process severely reduces the digestibility of the end product and its potential for value in lactating dairy cow diets. Elliott *et al.* (1994) found that resistance to ruminal and small intestinal lipolysis was a major factor contributing to the poor digestibility of highly saturated triglycerides contained in hydrogenated tallow.

2.2.7.2 Second generation fats

Calcium salts of fatty acids are the second generation of rumen inert fats. Palm oil, soybean oil, and other fat sources are hydrolysed and reacted with Ca to form salts, which increases the end product melting point. Fatty acids of Ca salts are stable in the rumen at pH less than 6.5. However, unsaturated fatty acids of (CaSFA) calcium soap fatty acid have been found to be extensively hydrogenated in the rumen. This indicated that dissociation occurred when the pH dropped below 6.5 after a meal or when the pH was manipulated *in vitro.* Wu and Palmquist (1991) observed that up to 55% of CaSFA were biohydrogenated. This indicated that CaSFA may not be as rumen inert as previously thought and may be deleterious to rumen fermentation and possibly to dry matter intake.

2.2.7.3 Third generation fats

The third generation free fatty acids are rumen inert fats. Rumen inert FFA are prehydrolysed, mostly hydrogenated, and purified during manufacturing. This form of rumen inert fat requires no further chemical modification by the cow prior to digestion. Free fatty acids usually have a lesser melting point than PHT or CaSFA and have the tendency to be less soluble in the rumen than fat supplements high in unsaturated fatty acids. Free fatty acids also have little or no negative effects on ruminal fermentation compared with fat sources high in unsaturated fatty acids. They have no any effect on dry matter intake of animal.

2.3 Level of Supplementation of Bypass Fat

In proximate analysis, dietary fats are expressed as ether extract, which includes true fats and ether soluble non-fatty acids substances. Up to 50% of forages and 20% of the grain ether extractable materials may be of non-fatty acids nature (Palmquist and Jenkins, 1980).

As per NRC (2001), dairy ration (mixture of cereals and forages) contains about 3% fat and the total dietary fat in ration should not exceed 6-7% of the DM.

Palmquist and Jenkins (1980) concluded that addition of 3-5% fat of the total ration DM has beneficial effect on the milk production; whereas decrease in production occurs, when the fat level exceeds 6% of the ration DM (Jenkins and Palmquist, 1984).

Palmquist (1991) suggested that the first 3% fat of the DM intake of the animal should be provided by oilseed sources and that in excess of 3% should be as bypass fat.

Sharma (2004) recommended that ration of the high producing animals should contain 4-6% fat, which should include fat from natural feed, oil seed and bypass fat in equal proportions.

Gowda *et al.* (2013) supplemented bypass fat @ 10g per kg milk production in lactating cows.

2.4 Effect of Protected Fat on Dry Matter Intake

Garg and Mehta (1998) study effect of bypass fat on feed intake, milk production and body condition score of Holstein cows. Bypass fat produced from rapeseed acid oil was incorporated in ration @ 500g/head/d for a period of six months after 7-12 days calving. The total DM intake (15.15+0.08 vs 16.10+0.12) was not affected by feeding bypass fat and control group

Sarwar *et al.* (2003) studied the effect of Beragfat T-300, a bypass fat, on dry matter intake and found that intake of DM was not affected by the supplementation of Bergafat in the diet of cows.

Lounglawan *et al.* (2007) studied the effect of feeding rumen-bypass fat on performance of dairy cows. They reported all cows consumed similar amount of total dry matter (DM).

Purushothaman *et al.* (2008) assessed the effect of feeding calcium soaps of palm oil fatty acids (bypass fat) on milk yield, milk composition and nutrient utilization in lactating crossbred cows. The average daily dry matter consumption in the various groups ranged from 13.1 to 13.6 kg and showed no significant difference among treatment groups.

Tyagi *et al.* (2009) reported increase (3.16 vs 3.41; kg/100 kg BW/d) in DM intake in dairy animals fed bypass fat.

Sirohi *et al.* (2010) concluded that dry matter intake (DMI) was higher for bypass fat supplementation treatment group compared to control group in medium producing cows. However, the values were non-significant.

Tyagi *et al.* (2010) evaluated the effects of supplementation with 2.5% (on DMI basis) bypass fat on milk production and reproductive performance of crossbred cows and reported non significant effect on dry matter intake of the cows.

Zhang *et al.* (2011) conducted an experiment to evaluate the effect of dietary fat supplementation with 1.5% palmitic acid, rapeseed oil or soybean oil, and fat powder on milk components and blood parameters of early-lactating cows under heat stress and reported that dry Matter Intake (DMI) was not affected by fat supplementation.

Naik (2013) reviewed that supplementation of bypass fat had no adverse effect on feed intake.

Mudgal *et al.* (2013) conducted an experiment on 12 crossbred cows on late lactation and their effect on dry matter intake and milk production. They reported that supplementation of berga fat did not have any effect on dry matter intake of cows.

Singh *et al.* (2014) find out the effect of prill fat feeding on milk production and hormonal changes in crossbred cows. The DMI of cows was non-significant ($P>0.05$) between the groups.

Sontakke *et al.* (2014) conducted in vivo study to observe the effect of feeding rice bran lyso-phospholipids (RBLP) and rumen protected fat (RPF) in 18 crossbred lactating Karan Fries (Holstein Friesian × Sahiwal) cows on DMI. The average DMI of cows was 12.14, 11.77 and 11.88 kg/d in control, rice bran lyso-phospholipids (RBLP) and rumen protected fat (RPF), respectively. There was non-significant difference ($P>0.05$) in DMI/100 kg body weight among 3 groups.

Ramteke *et al.* (2014) studied effect of bypass fat prepartum and during early lactation on productive performance in buffaloes. Twenty four pregnant buffaloes in their 2-4 lactation were selected on the basis of their milk production, fat per cent and stage of pregnancy for on farm trial of 150 days (30 days prepartum and 120 days post- partum) at an organized dairy farm of Gopalpura village of

Anand district. The buffaloes were divided into two dietary treatments i.e. T_1 (control) and T_2 (bypass fat). The buffaloes in control group (T_1) were fed as per farm feeding schedule (HMCM + green roughage + dry roughages) and those in bypass fat group (T_2) were reared on farm feeding schedule + by pass fat at 100 g/d for 30 days and 15g/kg milk yield per day for 120 days. The respective DMI during prepartum and post-partum periods were 8.00 and 7.79 kg/day and 11.42 and 11.41 kg/day in T_1 and T_2 groups, respectively, which were statistically similar in both the groups.

2.5 Effect of Protected Fat on Digestibility of Nutrients

Jenkins and Palmquist (1984) concluded that lowering of fat digestibility at higher level of supplementation may be due to the limited capacity of the small intestine to absorb the fat.

Grummer (1988) reported that the increase in the digestibility of the fat indicates that added fat is more digestible than the basal diet fat or fat supplementation dilutes the endogenous lipid secretions, resulting in more accurate estimate of the true lipid digestibility.

Palmquist (1991) reported that with the increase in fat intake, the apparent fat digestibility increases quadratically, while the true fat digestibility decreases linearly.

Schauff and Clark (1992) reported increase in the digestibility of CP, when Ca-LCFA was fed to the dairy animals.

Sarwar *et al.* (2003) studied the effect of Beragfat T-300, a bypass fat, on the production and composition of milk in primiparous crossbred cows in their early lactation and reported that digestibility of EE remained unaffected, whereas the digestibilities of DM and CP were reduced.

Naik *et al.* (2007) conducted experiment with 16 adult male buffaloes (421.4±10.4 kg BW) divided into four equal groups were offered conventional concentrate mixture (CM) or CM supplemented with 100, 200 or 300g PF along with wheat straw in roughage: concentrate ratio of 80:20 for 30 days. There was no difference in the digestibilities of nutrients except EE, which was improved ($P<0.05$) in PF200 and PF300 diets as compared to PF0 and PF100; while hemicellulose was improved in PF100 and PF200 over that of PF300 and concluded that protected fat did not have any effect when supplemented at 100g but showed improved EE digestibility at 200- 300 g in high wheat straw based diets

Purushothaman *et al.* (2008) selected twenty lactating crossbred cows yielding 10 to 15 litres of milk daily during mid lactation to assess the effect of feeding calcium soaps of palm oil fatty acids (bypass fat) at different level such as 2%, 4%, and 6% on milk yield, milk composition and nutrient utilization in lactating crossbred cows. There was no significant difference among different levels of bypass fat in digestibility of DM, OM, CP and CF, however, ether extract digestibility in cows at 2% and 4% bypass fat was significantly ($P<0.05$) higher than the control group.

Sirohi *et al.* (2010) studied the effect of bypass fat supplementation on nutrient utilization in 10 lactating crossbred cattle randomly divided into 2 groups on the basis of milk yield (14–15 kg/day). Cows were fed wheat straw, concentrate mixture and green maize fodder in the control group and additional 300 g bypass fat was given in treatment group. Experimental feeding was continued up to 90 days after 2 weeks of adaptation. The digestibility of nutrients except that of ether extract was comparable in both groups

Naik (2013_a) reviewed that supplementation of bypass fat had no effect on the feed intake and digestibility of nutrients.

Sontakke *et al.* (2014) conducted in vivo study to observe the effect of feeding rice bran lyso-phospholipids (RBLP) and rumen protected fat (RPF) in 18 crossbred lactating Karan Fries (Holstein Friesian × Sahiwal) cows. They were randomly allocated to 3 treatment groups having 6 animals in each group. Group 1 was supplemented with 2.5% mustard oil, group 2 with 6% RBLP and group 3 with 3% RPF in the concentrate mixture. All the 3 diets were isocaloric and isonitrogenous. All the cows were given roughage: concentrate mixture (60: 40) ration and roughage comprised of berseem fodder (30%) and wheat straw (30%) as per requirements. The digestibility of all the nutrients was similar among all groups except ether extract which was higher ($P<0.01$) in groups 2 and 3 as compared to group 1.

Yadav *et al.* (2015) observed that digestibility of nutrients was also similar in two groups except that digestibility of EE was higher ($P<0.01$) by feeding prilled fat to crossbred cows.

2.6 Effect of Protected Fat on Nutrient Intake

Sarwar *et al.* (2003) reported the intake of CP was not affected however; the EE intake was increased by the supplementation of Bergafat in the diet of cows.

Tyagi *et al.* (2009) reported decrease in the intake (kg) of DM (0.81 vs 0.78 and 0.82 vs 0.76); CP (0.12 vs 0.11; 0.12 vs 0.11) and TDN (0.52 vs 0.51; 0.52 vs 0.50) per kg of milk and FCM production in crossbred cows indicating better utilization of DM , CP and TDN due to bypass fat supplementation.

Sirohi *et al.* (2010) studied on lactating crossbred cattle (10) and randomly divided into 2 groups on the basis of milk yield (14–15 kg/day), day of calving

(~50 days) to see the effect of bypass fat supplementation on production performance. The DCPI (kg/d/animal) was 1.44+0.04 and 1.60+0.05 which was significantly differ and TDNI (kg/d/animal) was 8.33+ 0.32 and 9.24+0.24 which was similar in both group.

Naik *et al.* (2007b) reported that no effect on DCP and TDN intake by the supplementation of bypass fat to dairy animals.

Thakur and Shelke (2010) reported no effect on CP intake TDN by supplementation of bypass fat.

Ramteke *et al.* 2014 studied effect of bypass fat prepartum (30 days) and during early lactation (120 days) on productive performance in buffaloes. The buffaloes in control group (T_1) were fed as per farm feeding schedule (HMCM + green roughage + dry roughages) and those in bypass fat group (T_2) were reared on farm feeding schedule + bypass fat at 100 g/d for 30 days and 15g/kg milk yield per day for 120 days. DCPI during prepartum and post- partum was 760.79 g/d and 740.48 g/d, 888.69+0.01 and 888.31+4.2 respectively. The prepartum DCPI was statistically ($P<0.05$) higher in bypass fat supplemented group but during post-partum DCPI was statistically similar in both the groups. The average daily total digestible nutrients intake (TDNI) of experimental buffaloes during prepartum phase was 4.64+0.03 and 4.76+0.03 kg/head/day and during post-partum phase was 6.89±0.02 and 7.17±0.05 kg/head in T_1 and T_2 groups respectively. The TDNI in bypass fat group was significantly ($P<0.01$) higher in both the phases.

Sontakke *et al.* (2014) conducted in vivo study to observe the effect of feeding rice bran lyso-phospholipids (RBLP) and rumen protected fat (RPF) in 18 crossbred lactating Karan Fries (Holstein Friesian × Sahiwal) cows. DCPI and TDNI was significantly higher ($P<0.01$) in cows fed with 3% RPF than with 2.5% mustard oil on 100 kg body weight basis.

2.7 Effect of Protected Fat on Body Weight and Body Condition Score

Ferguson *et al.* (1994) reported body condition score (BCS) provides better estimate of body fat distribution than body weight.

Scot *et al.* (1994) reported increased body condition score at calving influenced treatment response of multiparous cows. Thinner cows produced more milk and less milk fat in response to additional dietary fat than fatter cows. Responses to rumen-inert fat by cows receiving high concentrations of dietary fat were marginal and were affected by body condition score at calving and by genetic potential.

Sklan *et al.* (1994) observed effect of supplemental bypass fat (Ca- LCFA) on change of BW is influenced by parity of the animal as loss of BW is more and longer lasting in primiparous cows than multiparous cows.

Garg and Mehta (1998) observed that the BSC of the cows improved due to bypass fat feeding indicating reduction in weight loss in the first quarter and helped gaining substantially after 90 days of feeding. The cows in control group lost more body weight than bypass fat feeding group.

Lounglawan *et al.* (2007) reported all cows lost similar live weight on feeding rumen bypass fat to dairy cows.

Naik *et al.* (2009b) reported better recovery in BW (-2.08 vs +14.13, kg) and BSC (-0.06 vs +0.02) in crossbred cows during early lactation in bypass fat supplemented group.

Thakur and Shelke (2009) reported that supplementation of calcium salts of soya acid oil fatty acids at 4% of DMI improved the average daily gain (553.10 vs 577.60, g) in Murrah buffalo calves owing to higher TDN intake (2.14 vs 2.42, kg/ d).

Wadhawa *et al.* (2012) assessed the effect of strategic supplementation of bypass fat (BPF) on the body weight change of high yielding crossbred cows. The body weight of the animals fed with adlib control diet + 150-200 g of calcium salts of rice bran fatty acid oil (BPF) was improved considerably as compared to control group, but was not affected by stage of lactation.

Gowda *et al.* (2012) reported less body weight loss in cows fed protected fat with a moderate effect the regain of body weight was much quicker as compared to cows maintained on existing farmer's feeding schedule under field condition.

Naik (2013a) reviewed that the milk yield is increased by 5.5-24.0% along with the improvement in post partum recovery of the body weight and body condition score of the dairy animals.

Mathew and Zachariah (2014) evaluated effect of Rumen Escape Fat (REF) supplementation on early lactation in crossbred cows and reported that animals in both treatment groups that were supplemented with bypass fat showed yielded better VABCS than control group. Supplementation of 100 grams of bypass fat was found to be effective in preventing weight loss during initial 90 days of lactation.

Singh *et al.* (2014) find out the effect of prill fat feeding on milk production and hormonal changes in crossbred cows.The body weight of cows was non-significant ($P>0.05$) between the groups.The decline in BCS ($P<0.01$) was more in control group cows than the prill fat cows group.

Kirovski *et al.* (2015) conducted study to test the effect of rumen-inert fat supplement of palm oil on milk production, milk composition, rumen characteristics, and metabolic variables of early lactating dairy cows. The loss in body condition was significantly lower in the group fed palm oil than in the control group. Body condition scores at day 58 of control cows were significantly lower than those at days 30 and 86 of lactation ($P < 0.05$, respectively) and were significantly lower than the value obtained at the same time period in the experimental group ($P< 0.01$).

Yadav *et al.* (2015) assessed effect of prilled fat supplementation on certain plasma metabolites and milk production in Karan Fries cows. Body condition score of treatment cows was more ($P<0.05$) in prilled fat supplementation group comparison to control group.

2.8 Effect of Protected Fat on Milk Yield

Grummer (1988) concluded that transfer efficiency of plasma fatty acids to mammary tissue decreases as lactation progresses; therefore, increase in production is maximal during early and peak lactation than mid or late lactation.

West and Hill (1990) reported there is no significant interaction with breed of cow and effect of supplemental bypass fat (Ca-LCFA) on milk yield which tends to be greater in Holstein cows than Jersey cows.

Sklan *et al.* (1991) reported that due to the higher energy intake, more efficient use of fat by mammary gland and enhancement of tissue mobilization, milk yield and FCM yield increased in early and peak lactation.

Schauff and Clark (1992) reported increase in FCM yield of lactating cows, when Ca-LCFA was supplemented up to 6% of the dietary DM, but, it decreased at 9% of the dietary DM.

Chouinard *et al.* (1997) reported higher level of supplementation of Ca-LCFA, increase in milk yield is quadratic, as it interferes with the digestion of other nutrients and impairs benefits of the supplemental fat on energy utilization.

Garg and Mehta (1998) study effect of bypass fat on feed intake, milk production and body condition score of Holstein cows. Bypass fat produced from rapeseed acid oil was incorporated in ration @ 500g/head/d for a period of six months after 7-12 days calving. Feeding bypass fat significantly ($P<0.05$) improved milk yield and 4% fat corrected milk. Also reported that bypass fat feeding had maximum effect on milk yield during the first quarter of the lactation, when feed intake is usually low and the effect was less prominent as lactation advanced, probably due to the DM intake start increasing after 6-8 weeks of calving.

Further, production of the lactating animal is dependent on the amount of bypass fat supplemented in the ration.

Chouinard *et al.* (1998) reported that milk production of dairy cows was influenced by the fatty acids composition of the Ca salts, which was further dependent upon the lactational stage of the animal. With increase in the unsaturation of the dominant fatty acids in Ca salts, milk yield increased linearly in early lactation.

Sarwar *et al.* (2003) studied the effect of Beragfat T-300, a bypass fat, on the production and composition of milk, four primiparous crossbred cows in their early lactation were used in a 4×4 Latin Square Design. Each period was of 30 days including 15 days of adjustment period. The diets were formulated to contain 0, 2.5, 3.5 and 4.5% of Bergafat and were isonitrogenous and isoenergetic. Milk yield lit per day remained unaltered (12.7, 12.7, 13.3 and 12.9) while 4% FCM yield lit per day (12.2, 14.6, 15.4 and 15.4) significantly increased as a result of adding Bergafat in the daily ration.

Lounglawan *et al.* (2007) reported no significant differences in milk yield and 3.5% FCM on feeding rumen bypass fat to dairy cows.

Purushothaman *et al.* (2008) selected twenty lactating crossbred cows yielding 10 to 15 litres of milk daily during mid lactation and assessed the effect of feeding calcium soaps of palm oil fatty acids (bypass fat) at different levels such as 2%, 4%, and 6% on milk yield, milk composition and nutrient utilization in lactating crossbred cows. The animals in groups 1 (control), 2, 3 and 4 were fed concentrate mixture containing 0 (no bypass fat), 2, 4 and 6% bypass fat, respectively. The average daily milk yield was significantly ($p<0.05$) higher in cows of group 3 (13.6 kg) that were fed 4% bypass fat than in cows of groups 1 (11.7 kg) and 2 (12.0 kg). But there was no significant difference in milk yield between the cows of groups 3 and 4 (12.7 kg), as the 4% bypass fat supplementation showed a significant ($P<0.05$) increase in milk yield (19.0%) as compared to control. Similarly, a significant ($P<0.05$) increase in fat yield, 4% FCM yield and SNF yield was observed for the cows in group 3 (4% bypass fat) than control, 2% and 6%.

Barley and Baghel (2009) studied effect of bypass fat supplementation on milk yield, fat percentage and serum triglyceride levels of Murrah buffaloes. Supplementation of bypass fat (100g/ days) resulted in immediate increase in milk yield during the first week and sustained the increasing trend in subsequent weeks of lactation. This trend can be attributed to high energy supplementation by way of bypass fat which brings the animal out of negative energy balance.

Bhosale *et al.* (2010) reported that supplemented bypass fat (100g/h/day) to lactating buffalo significantly increased milk yield.

Sirohi *et al.* (2010) concluded that bypass fat supplementation @300 g/d/animal in medium producing crossbred cows significantly increased the milk production and fat corrected milk yield up to 15.61 and 24.01%, respectively, over the control group.

Tyagi *et al.* (2010) evaluated the effects of supplementation of bypass fat on milk production performance of crossbred cows. The cows under treatment groups were supplemented bypass fat 40 days prepartum to 90 days postpartum and carry over effect of supplementation on milk production was monitored up to 210 days of lactation. The average milk yield (210 days) in supplemented bypass fat group was higher ($P<0.05$) than that of control group (18.65 vs 17.57 kg/day). Similarly, the milk yield during the carry over period (90 days) was higher ($P< 0.05$) in supplemented bypass fat group than that of control group (14.81 vs 13.79 kg/day).

Zhang *et al.* (2011) conducted an experiment to evaluate the effect of dietary fat supplementation on milk components and blood parameters of early-lactating cows under heat stress. Cows in group 1 were fed basal diet without dietary fat (T_1) supplementation (control). The other four experimental groups were fed 1.5% palmitic acid (T_2), rapeseed oil (T_3) or soybean oil (T_4), and 1.8% fat powder (T_5), respectively. The milk yield increased for cows fed supplemental fat. In the control group there was 1.42 kg reduction of milk yield, the decrease being 6.7%. In soybean oil group (T_2) milk yield decrease was by 0.42 kg, which is by 2.02%. On the other hand, in fat powder group (T_3) an increase by 0.36 kg was noted, the increment accounting for 1.85% in palmitic acid group (T_4) there was an increase in milk production by 2.50 kg, the increase being 13.3%; and, in rapeseed oil group (T_5) the increased in milk yield was by 0.08 kg, accounting for 0.41% enhancement.

Wadhawa *et al.* (2012) assessed the effect of strategic supplementation of bypass fat (BPF) with 150 - 200 g of calcium salts of rice bran fatty acid oil on the performance of high yielding crossbred cows. The average daily milk yield was improved by 1.13 kg/d (21.55 vs. 20.42 kg/d) in BPF supplemented group as compared to control group.

Ranjan *et al.* (2012) studied the effect of dietary supplementation of bypass fat on productive performance and blood biochemical profile of mid lactating Murrah buffaloes (Bubalus bubalis). The milk yield was not influenced by supplemental bypass fat. However, fat-corrected milk (6.5 %) yield was higher ($P<0.05$) in supplementation of bypass fat at 1.4% (200 g/ day) (14.21) than that of control group (9.83) and similar with 0.7 % (100g/day) (11.05)..

Qureshi and Tawheed (2012) conducted study to investigate the effect of protected palm fats feeding on milk fatty acids profiles of crossbred cows. A total of 15 of crossbred and 15 of Holstein Friesian cows were selected and protected palm fats were supplemented as: PF-0, PF-25, PF-50, PF-100 and; PF-150; the number representing the quantity (g) of fats/day. Highest milk yield (13.31±0.81 kg/day) by supplementation of protected fat in PF150 group, followed by 12.66±0.79 kg/day in PF100 supplemented group ($P<0.05$).

Gowda *et al.* (2013) studied the effect of protected fat supplementation to high yielding dairy cows in field condition. Twelve numbers of high yielding crossbred (Holstein Frisian) dairy cows in their 2-5th lactation maintained by farmers were selected based on their previous lactation yield. The average milk yield was significantly ($P<0.01$) higher with large effect size (Av. 19.1 vs 17.8 lit/cow/day) in cows supplemented with protected fat.

Naik (2013a) reviewed that the milk yield increased by 5.5-24.0% of the dairy animals by feeding bypass fat.

Mudgal *et al.* (2013) conducted an experiment on supplementation of berga fat 12 crossbred cows (Milk production 6.52± 11.15 liters) in late lactation and their effect on dry matter intake and milk production. They reported that supplementation of berga fat did not have any effect on production performance due to treatment, which indicated that in late stage of lactation when production was on declining trend and which was also supported by reduced nutrients demand had reduced dry matter intake in the cows. They concluded that positive effects of bypass fat supplementation may only be observed in the stage of early lactation and due to reduced demand of nutrients the advantage of feeding bypass fat was not observed in later stage of lactation.

Mathew and Zachariah (2014) evaluated effect of Rumen Escape Fat (REF) supplementation on early lactation in crossbred cows. Animals in control group received no supplementation in diet of rumen escape fat popularly called bypass fat whereas animals allotted to treatment 1 group received 100 grams supplementation and those in treatment 2 group received daily supplementation of 150 grams in diet from 7th day from parturition till end of trial. Average milk yield in litres of animals were improved in all groups from 9.14 litres on 7th day of parturition to 14.27 litres, 15.45 litres and 15.09 litres in control, treatment I and treatment II, respectively in trial period that lasted six fortnights. Increase in milk production in all groups might be explained by higher feed intake towards end of trial when transient anorexic phase due to involuting uterus tides over.

Singh *et al.* (2014) find out the effect of prill fat feeding on milk production and hormonal changes in crossbred cows. Milk yields during supplemental of 90 days increased @ 0.50 kg/ d/cow in the supplemented with Prill Fat over the control. However, fat corrected milk, (FCM) and fat corrected milk energy was similar. The Energy corrected milk yield (kg/d) was higher ($P<0.05$) in Prill Fat feeding in comparison to control. The milk yield of cows remained higher ($P<0.05$) even after the withdrawal of prill fat feeding @0.71kg/d in comparison to control.

Sontakke *et al.* (2014) conducted in vivo study to observe the effect of feeding rice bran lyso-phospholipids (RBLP) and rumen protected fat (RPF) in 18 crossbred lactating Karan Fries (Holstein Friesian × Sahiwal) cows. The daily milk yield of lactating cows in groups 1, 2 and 3 was 12.54, 12.90 and 12.84 kg, respectively showing no significant difference ($P>0.05$) between all the 3 groups. Cows in all the 3 groups maintained their milk production throughout the experimental period of 90 days even though the experiment was started at about 100 days of lactation which might be due to the inclusion of fat source in one or the other form. Overall mean 4% FCM yield was 13.65, 14.47 and 14.25 kg/d in the 3 respective groups showing significantly higher ($P<0.05$) milk yield in groups 2 and 3 than group 1, which was due to significant increase in fat % in the milk of this group.

Ramteke *et al.* (2014) studied effect of bypass fat prepartum and during early lactation on productive performance in buffaloes . Twenty four pregnant buffaloes in their 2-4 lactation were selected on the basis of their milk production, fat per cent and stage of pregnancy for on farm trial of 150 days (30 days prepartum and 120 days post- partum) at an organized dairy farm of Gopalpura village of Anand district. The buffaloes were divided into two dietary treatments i.e. T_1 (control) and T_2 (bypass fat). The buffaloes in control group (T_1) were fed as per farm feeding schedule (HMCM + green roughage + dry roughages) and those in bypass fat group (T_2) were reared on farm feeding schedule + bypass fat at 100 g/d for 30 days and 15g/kg milk yield per day for 120 days. The whole milk yield (kg/d) and FCM was statistically ($P<0.05$) higher in bypass fat supplemented group than control. Similar trend was also observed in case of SCM and ECM yield.

Kirovski *et al.* (2015) conducted study to test the effect of rumen-inert fat supplement of palm oil on milk production, milk composition, rumen characteristics, and metabolic variables of early lactating dairy cows. For this purpose, 24 Holstein-Friesian cows were divided into two equal groups and fed a corn silage-based diet, without palm oil supplementation (control) or with 300 g palm oil (Palm Fat 99, Noack & Co. GmbH, Vienna, Austria) per cow for 8 weeks starting from day 30 after parturition. Compared with the control, palm oil supplementation resulted in an increase of the average milk yield. Mean daily milk yields at days 58 and 86 of lactation were higher than those in the control cows but the difference was not significant.

Yadav *et al.* (2015) assessed effect of prilled fat supplementation on milk production in Karan Fries cows. Cows fed prilled fat group (14.34 ± 0.16) produced significantly ($P<0.05$) more milk as compare to control (12.90 ± 0.19) and also 4% FCM produced significantly more milk in prilled fat group (15.65 ± 0.18) than control (13.06 ± 0.21).

2.9 Effect of Protected Fat on Milk Composition

Palmquist and Moser (1981) reported dietary protein should be increased, when fat supplemented diets are fed to animals. They reported that dietary fat impairs amino acids transport to mammary gland and decreases milk protein synthesis by inducing insulin resistance, if protein or amino acids are inadequate, lipoprotein synthesis may be decreased. Consequently, fat and protein transport to mammary gland is reduced leading to decrease in milk protein percentage, which usually occurs and decrease in milk fat percentage.

Ferguson *et al.* (1989) observed supplementation of Ca salts in diet of lactating cows increases milk yield without changing milk composition.

West and Hill (1990) reported that among all components of milk, fat content is most sensitive to the dietary changes. Similar to milk yield, although there is no significant interaction with breed of cow, effect of supplementation of Ca-LCFA on milk composition tends to be greater in Holstein cows than Jersey cows on supplementation of bypass fat to lactating animals. Also they observed effect of supplemental Ca-LCFA on milk protein content is not influenced by the breed of the animal but is influenced by the parity and stage of lactation of the animal. Decline in milk protein content is reported both in multiparous cows.

DePeters and Cantt (1992) expressed depressed milk protein percentage is related to the dilution of milk protein as higher milk volume synthesized is not synchronized with uptake of amino acids by the mammary gland.

Chouinard *et al.* (1998) concluded that supplementation of bypass fat (Ca-LCFA) has negative effect on the milk protein percentage; an overall effect of -0.12 percentage unit. Also decreased milk fat percentage on supplementation of bypass fat to lactating animals.

Sarwar *et al.* (2003) studied the effect of Beragfat T-300, a by pass fat, on the production and composition of milk in early lactation cows. The diets were formulated to contain 0, 2.5, 3.5 and 4.5% of Bergafat. Milk fat percentage was highest (5.3%) at 4.5% added Bergafat and lowest (3.7%) in the control. The milk protein per cent was highest @ 4.5% berga fat group (3.4) and lowest in all other group (3.3) which was non significant difference among treatment.

Lounglawan *et al.* (2007) reported non significant difference in fat, protein, lactose, solid not fat and total solid percentage in milk feeding rumen-bypass fat to dairy cows.

Purushothaman *et al.* (2008) reported that bypass fat feeding did not have any effect on milk composition of mid lactating crossbred cows.

Barley and Baghel (2009) studied effect of bypass fat supplementation on milk yield, fat percentage and serum triglyceride levels of Murrah buffaloes. The treatment group was offered bypass fat supplement 100 gs once in a day/animal. The trial was conducted for 45 days. During the first week of trial, the highest fat percentage was recorded for the bypass fat supplemented group, which was significantly higher than the control group. The trend of higher fat percentage in the bypass fat supplemented group continued during second third through the fourth week.

Naik *et al.* (2009b) reported not alteration in milk fat, milk protein, lactose and SNF on supplementation of bypass fat to lactating animals but addition of bypass fat in diet generally increases the total milk fat yield due to increase in the milk production, the total milk protein yield, SNF yield, lactose yield and milk fat yield was increased due to the increase in milk production.

Tyagi *et al.* (2009a) studied effect of feeding bypass fat supplement during the first quarter of lactation on milk production on crossbred cows reported non significant difference in milk composition in treatment group fed with 2.5% bypass fat of total DMI.

Thakur and Shelke (2010) reported that milk fat percentage was increased on supplementation of bypass fat to lactating animals.

Sirohi *et al.* (2010) reported milk fat and total solids content were improved significantly in treatment group, whereas milk protein and solids-not-fat remained unaffected in both groups by supplementing bypass fat on production performance of lactating crossbred cows.

Zhang *et al.* (2011) conducted an experiment to evaluate the effect of dietary fat supplementation on milk components and blood parameters of early-lactating cows under heat stress. Cows in group 1 were fed basal diet without dietary fat (T_1) supplementation (control). The other four experimental groups were fed 1.5% palmitic acid (T_2), rapeseed oil (T_3) or soybean oil (T_4), and 1.8% fat powder (T_5), respectively. Milk fat increased significantly by fat supplementation ($P<0.05$), while milk protein and lactose were not significantly altered by fat supplementation ($P>0.05$).

Wadhawa *et al.* (2012) concluded that supplementation of 150–200 g bypass fat/d improved the milk yield, its composition with an increase in protein, lactose and SNF content in milk, without effecting milk fat content as compared to control group.

Ranjan *et al.* (2012) studied investigated the effect of dietary supplementation of bypass fat on productive performance and of lactating Murrah buffaloes (Bubalus bubalis). The milk composition were not influenced by supplemental bypass fat.

Mathew and Zachariah (2014) evaluated effect of Rumen Escape Fat (REF) supplementation on milk fat percentage of early lactating crossbred cows. Animals in control group received no supplementation in diet of rumen escape fat popularly called bypass fat whereas animals allotted to treatment 1 group received 100 grams supplementation and those in treatment 2 group received daily supplementation of 150 grams in diet from 7th day from parturition till end of trial. Average milk fat percentage of animals were improved in all groups from 3.22 per cent on 7th day of parturition to 3.65 per cent, 3.70 per cent and 3.75 per cent in control, treatment 1 and treatment 2, respectively in the trial period that lasted six fortnights.

Singh *et al.* (2014) find out the effect of prill fat feeding on milk production and hormonal changes in crossbred cows. Milk fat, protein and lactose were similar in both the groups.

Sontakke *et al* (2014) conducted in vivo study to observe the effect of feeding rice bran lyso-phospholipids (RBLP) and rumen protected fat (RPF) in crossbred lactating Karan Fries (Holstein Friesian × Sahiwal) cows. The protein, lactose, total solids, and solids-not-fat were not affected by dietary treatments except fat.

Ramteke *et al.* (2014) studied effect of bypass fat prepartum and during early lactation on productive performance in buffaloes for on farm trial of 150 days (30 days prepartum and 120 days post- partum) at an organized dairy farm of Gopalpura village of Anand district. The content of fat % was statistically higher in bypass fat supplemented group. The milk fat per cent was found to be significantly ($P<0.01$) higher in bypass fat supplemented group due to more availability of fatty acids for absorption in intestine due to protection of fat and these fatty acids are directly incorporated in milk fat after absorption from intestine, leading to increase in milk fat. However, content of SNF % was significantly reduced in bypass fat supplemented group.

Kirovski *et al.* (2015) conducted study to test the effect of rumen-inert fat supplement of 300 g palm oil on milk production, milk composition, rumen characteristics, and metabolic variables of early lactating dairy cows. Average milk fat contents at day 58 were 2.93 ± 0.22% (control cows) and 2.46 ± 0.20%

(experimental cows), while at day 86, average milk fat contents were 2.59 ± 0.19% in the control cows and 3.25 ± 0.22% in the experimental one. At day 86, fat contents in the experimental cows were significantly higher than those at day 56 in the same group ($P< 0.01$) as well as than those at day 86 in the control group ($P< 0.05$). Average milk protein contents at day 56 were 3.05 ± 0.03% and 3.01 ± 0.03% in the control and the experimental cows, respectively. At day 86 of lactation, average milk protein contents were 3.00± 0.04% (control group) and 3.07± 0.03% (experimental group) which was non significant difference.

Yadav *et al.* (2015) assess the effect of prilled fat supplementation on certain plasma metabolites and milk production in Karan Fries cows. The animals in control group were fed as per requirements while treatment group also received prilled fat @ 75 g/cow/day in addition to control from 45 days prepartum till parturition. After parturition the treatment group received prilled fat @ 150g/cow/day till day 70th of lactation. Cows in treatment group produced more milk with higher fat and SNF contents ($P<0.01$).

2.10 Effect of Protected Fat on Fatty Acid Profile

Chouinard *et al.* (1997) reported that the proportions of odd numbered FA (C15:0, C17:0) of milk fat decreases because ruminal bacteria, which are major source of odd chain fatty acids of milk fat prefers to use preformed fatty acids and de novo fatty acids synthesis is reduced.

Mishra *et al.* (2004) reported supplementation of Ca-LCFA in the diet of lactating cows generally decreases the proportions of short and medium chain saturated fatty acids (C6:0 to C16:0) of milk fat due to reduction in de novo FA synthesis in mammary gland and increase in proportions of LCFA (C18:1, C18:2, C18:3) due to increased uptake of preformed LCFA from blood

Lounglawan *et al.* (2007) reported rumen-bypass fat supplementation significantly ($P<0.05$) reduced C4:0 but increased C12:0 and C14:0 fatty acids of the cows' milk

Purushothaman *et al.* (2008) assessed the effect of feeding calcium soaps of palm oil fatty acids (bypass fat) on milk yield, milk composition and nutrient utilization in lactating crossbred cows. The supplementation with calcium salts of palm oil fatty acid reduced the proportion of caproic, caprylic and capric acids and significantly ($P<0.01$) increased the concentration of palmitic, oleic, stearic, linoleic and linolenic acids in milk fat with increase in level of bypass fat supplementation.

Tyagi *et al.* (2009a) observed during the total lactation (early, mid and late) period, there was increase in the total USFA (32.01 vs 39.22), LCFA (75.61 vs

77.17) and MUFA (29.68 vs 33.53) and decrease in the total SFA (63.28 vs 54.02) as percentage of the total fatty acids of milk due to supplementation of bypass fat in the diet of dairy cows.

Tyagi *et al.* (2010) studied effect of feeding bypass fat supplement during the first quarter of lactation on milk production and milk composition. The cows of treatment group were supplemented bypass fat 40 days pre partum till 90 days post partum and carry over effect of supplementation on milk production was monitored up to 210 days of lactation reported total unsaturated fatty acids (as % of total FA) in control group were 32.49, 32.03 and 31.52 per cent in early, mid and late lactation, whereas corresponding values in treatment group were 41.36, 38.00 and 38.29, respectively showing an increase of 27.3, 18.6 and 21.5 in treatment group over that of control group. The total long chain fatty acids (LCFA) and monounsaturated fatty acids (MUFA) content were higher in treatment group as compared to that of control group throughout the experimental period.

Qureshi and Tawheed (2012) conducted study to investigate the effect of protected palm fats feeding on milk fatty acids profiles of crossbred cows. A total of 15 of crossbred and 15 of Holstein Friesian cows were selected and protected palm fats were supplemented as: PF-0, PF-25, PF-50, PF-100 and; PF-150; the number representing the quantity (g) of fats/day. SFA was significantly ($P<0.05$) decreased from 70.80 to 67.45 g/100g while MUFA and PUFA increased with the increasing supplementation. It appears that hypercholestermic properties of the milk were reduced and cardio-protective properties were enhanced by feeding protected palm fats.

Sontakke *et al.* (2014) conducted in vivo study to observe the effect of feeding rice bran lyso-phospholipids (RBLP) and rumen protected fat (RPF) in 18 crossbred lactating Karan Fries (Holstein Friesian × Sahiwal) cows. They were randomly allocated to 3 treatment groups having 6 animals in each group. Group 1 was supplemented with 2.5% mustard oil, group 2 with 6% RBLP and group 3 with 3% RPF in the concentrate mixture. The proportion of total unsaturated fatty acid content of milk in groups 2 and 3 was higher ($P<0.05$) to that of group 1. The C18:2 fatty acid in the milk was increased in group 2 as compared to group 1 and was almost similar to group 3. The saturated fatty acid content in milk of group 1 was higher in comparison to that of groups 2 and 3 cows. Total long chain fatty acids (TLCFA) in groups 1, 2 and 3 were showing no difference in different treatment groups.

2.11 Effect of Protected Fat on Reproduction

Sklan *et al.* (1991) observed feeding Ca-LCFA increases pregnancy rate and reduces open days.

Sklan *et al.* (1994) presented several hypotheses suggested regarding role of the fatty acids on reproductive performance of dairy animals. These include (i) improved energy balance results in an earlier return to post-partum ovarian cycling; (ii) increase linoleic acid may provide increase PGF2 alpha and stimulate return to ovarian cycling and improve follicular recruitment; and (iii) increase in progesterone secretion either from improved energy balance or from altered lipoprotein composition from dietary fat improves fertility.

Scott *et al.* (1995) reported changes in reproductive performance associated with fat supplementation related to magnitude of the milk response of the fat supplementation.

Garg and Mehta (1998) find out that due to bypass fat feeding, the average period for conception after calving was reduced (118 vs 92 days) in cows.

Naik *et al.* (2009b) reported that when bypass fat was included in diet of crossbred cows, the number of artificial inseminations required per conception was reduced (1.4 vs 1.2), indicating better reproductive performance of animals.

Tyagi *et al.* (2010) evaluated the effects of supplementation of bypass fat on milk production and reproductive performance of multiparous crossbred cows. Nineteen multiparous crossbred cows (2–4 lactation) were divided in two groups on the basis of most probable production ability (MPPA). The animals in group 1 (nine cows, MPPA 3,441.32 kg, control group G_1) were fed chaffed wheat straw, chopped green maize, and concentrate mixture as per requirements while the animals in group 2 (10 cows, MPPA 3,457.2 kg, treatment group G_2) were fed the same ration supplemented with 2.5% bypass fat (on DMI basis). The cows of G_2 were supplemented bypass fat 40 days prepartum and studied effect on calving and birth weight of calves. Average birth weights of the calves were 24.94 and 27.95 kg in G_1 and G_2, respectively. The calving per cent in G_1 (88.88%) was lower than that of G_2 (100%). The time taken for expulsion of fetal membranes was decreased ($P<0.05$) by 5.4 h in G_2 compared to G_1. Days required for involution of uterus was less ($P<0.05$) in G_2 (35.40 days) than that of G_1 (49.44 days). Less number of cases of retention of fetal membranes (RFM) and metritis were observed in G_2 as compared to that of G_1.

Shelke *et al.* (2012) reported that supplementation of protected fat and protein during early lactation improved the reproductive performance in Murrah buffaloes along with increase in the milk production and its persistency.

Naik (2013a) reviewed that the milk yield is increased by 5.5-24.0% along with the improvement in reproductive performance of the dairy animals.

Qureshi and Tawheed (2012) reported that feeding protected fat associated with increased milk yield and progesterone level reflecting better fertility and productivity and suggested that 150 g/day palm protected fat may be supplemented in early lactation for maximum yield, better reproductive performance and healthier milk.

Gowda *et al.* (2013) reported significantly ($P<0.05$) better reproductive performance with cows fed protected fat. The time required for first heat appearance was reduced ($P<0.05$) by 18 days in bypass fat feeding as compared to control. The number of artificial insemination (AI) per conception was reduced significantly ($P<0.05$) in bypass fat feeding group (2.4 vs 1.8). Consequently, the number of days open was also reduced ($P<0.05$) in treatment group by 25 days than that control group, indicating superior reproductive performance in cows of bypass fat feeding. The statistical significance for these attributes was large (L) to very large (VL), confirming the beneficial effects of protected fat supplementation during the initial period of lactation.

Bahram Rahbar *et al.* (2014) concluded that fat is recommended to be incorporated into farm animal diets at moderate amounts. Feeding fat to cattle generally improved establishment and maintenance of pregnancy. Potential improvements in fertility of cows caused by fat feeding have generally been associated with enhanced follicle development postpartum, increased diameter of the ovulatory follicle, increased progesterone concentrations during the luteal phase of cycle, altered uterine/ embryo cross-talk by modulating prostoglandin synthesis, and improved oocyte and embryo quality.

2.12 Effect of Protected Fat on Various Blood Parameters

Barley and Baghel (2009) studied effect of bypass fat supplementation on serum triglyceride (mg/dl) levels of Murrah buffaloes and divided in two groups of 20 each a supplemented and a non- supplemented control group. The feed ingredients and fodder offered to them was unchanged during the period of trial and the treatment group was offered bypass fat supplement 100 gs once in a day/animal. The trial was conducted for 45 days. During early lactation, the pretrial serum triglyceride level ranged from 31.02+0.58 to 31.18+0.77 during the first week and from 31.07+0.61 to 31.34+0.33 during the second week. Non-significant differences were observed amongst the treatments. During the trial, the serum triglyceride of control animals ranged from 31.16+0.70 (I week) to 31.45+0.70 (IV Week), while the serum triglyceride level of bypass fat supplemented animals ranged from 37.23+0.70 (I Week) to 41.60+0.17 (IV Week). A significant difference was observed in the serum triglyceride levels of animals fed bypass fat compared to control animals. Immediate increase in serum triglyceride in the supplemented animals compared to the control animals.

Tyagi *et al.* (2009a) reported NEFA concentration in blood was 106.54 and 124.12 mg/dl and Plasma triglyceride concentrations were 18.78 and 18.15 mg/dl in control and treatment group, respectively. The average blood glucose concentrations were similar in both the groups being 58.64 and 57.82 mg/ dl in control and treatment group, respectively. Average BUN value was lower in treatment group but the variation was statistically non significant.

Tyagi *et al.* (2010) studied the effects of supplementation of bypass fat on blood parameter of crossbred cows and reported non significant difference in plasma cholesterol concentrations.

Zhang *et al.* (2011) conducted an experiment to evaluate the effect of dietary fat supplementation on milk components and blood parameters of early-lactating cows under heat stress. Cows in group 1 were fed basal diet without dietary fat (T_1) supplementation (control). The other four experimental groups were fed 1.5% palmitic acid (T_2), rapeseed oil (T_3) or soybean oil (T_4), and 1.8% fat powder (T_5), respectively. The concentration of plasma glucose was increased by fed supplemental fat ($P<0.05$) while that of plasma urea nitrogen was decreased by fed supplemental fat ($P<0.05$). Furthermore, there were significant increases of total triglycerides, total cholesterol and high density lipoprotein (HDL) ($P<0.05$), but low-density lipoprotein (LDL) did not differ significantly by fat supplementation ($P>0.05$).

Wadhawa *et al.* (2012) assessed the effect of strategic supplementation of bypass fat (BPF) on the blood profile of high yielding crossbred cows. Crossbred cows (15) in early lactation, producing more than 10 kg milk/d were used. The animals in the experimental group were offered *ad lib.* control diet, supplemented daily with 150– 200g BPF i.e. calcium salts of rice bran fatty acid oil. The BPF supplementation did not have any significant impact on blood profile except an increase in triglyceride (TG) level also the blood uric acid levels are increased ($P<0.05$) in the animals fed bypass fat.

Ranjan *et al.* (2012) investigated the effect of dietary supplementation of bypass fat on productive performance and blood biochemical profile of lactating Murrah buffaloes *(Bubalus bubalis*). The serum cholesterol level was higher ($P <0.01$) in bypass fat-supplemented group (T_2) and (T_3) of animals. Serum high density lipoprotein (HDL) cholesterol (good cholesterol) level was more ($P <0.05$) than LDL cholesterol (bad cholesterol) level with higher dose of bypass fat in (T_3) than (T_2).

Naik (2013a) reviewed that supplementation of bypass fat had no effect on blood glucose, BUN, NEFA, protein, albumin, globulin, creatinine and cholesterol level of the dairy animals.

Singh *et al.* (2014) find out the effect of prill fat feeding on milk production and hormonal changes in crossbred cows. Plasma NEFA (0.20+0.03 vs 0.14+0.03) decreased (P<0.05) and glucose (57.36+2.36 vs 55.19+3.06) varied non-significantly between the groups prill fat supplemented cows over the control cows. Plasma cholesterol (169.37+11.30 vs 176.94+12.16) increased (P<0.01) and HDL decreased in prill fat supplemented cows over the control cows, but the values did not reach statistically significant. Plasma triglycerides (31.47+2.04 vs 25.96+1.24) levels were not affected (P>0.05) in both the groups.

Kirovski *et al.* (2015) conducted study to test the effect of rumen-inert fat supplement of palm oil on milk production, milk composition, rumen characteristics, and metabolic variables of early lactating dairy cows. Blood samples are taken three times during experiment at days 30, 58, and 86 of lactation. Blood samples were tested for total protein, albumin, urea, tryglicerides, glucose, cholesterol, total bilirubin, beta-hydroxybutyrate, Ca, and P. Results show that, urea and glucose concentrations were significantly lower in the palm oil-supplemented compared to the control group (P < 0.05 and P<0.01, respectively); at day 58, Ca concentration was significantly higher in the palm oil-supplemented compared to the control group (P<0.05); and at day 86, cholesterol concentration was significantly higher in the experimental compared to the control group (P< 0.01). Total protein, albumin, urea and tryglicerides are not effected by palm oil-supplemented.

Yadav *et al.* (2015) studied the effect of prilled fat supplementation on certain plasma metabolites and milk production in Karan Fries cows. The animals in control group were fed as per requirements while treatment group also received prilled fat @ 75g/cow/day in addition to control from 45 days prepartum till parturition. After parturition the treatment group received prilled fat @ 150g/cow/day till day 70th of lactation. The concentration of plasma NEFA (P<0.01) and total cholesterol was lower (P<0.05) in treated cows than the control group.

2.13 Effect of Protected Fat on Economics of Milk Production

Naik *et al.* (2009b) reported that feeding of the indigenously prepared bypass fat to dairy animals has shown to give additional profit of Rs. 34.50/- per cow per day besides improvement in reproductive performance and health of the animals.

Parnerkar *et al.* (2011) reported that feeding of the indigenously prepared bypass fat to buffalo has shown to give additional profit Rs. 39.66/- per day.

Gowda *et al.* (2013) reported that on feeding of protected fat resulted in a net profit of Rs. 11.6 per cow per day due to higher milk production and concluded that protected fat supplementation to cows maintained on exiting feeding practices at field condition improved the milk production and reproductive efficiency in dairy cattle.

Naik (2013b) observed that feeding of the indigenously prepared bypass fat to lactating dairy animals has shown to give additional benefit of Rs. 12-40/- per animal per day.

Singh *et al.* (2014) observed that prill fat feeding incurred extra cost of Rs.8/day/animal during the experimental period and generated additional income of Rs.50/day/animal by increased milk and fat yield. The total income return from the experimental prill fat fed cows was significantly more ($P<0.05$) than control.

Sontakke *et al.* (2014) conducted in vivo study to observe the effect of feeding rice bran lyso-phospholipids (RBLP) and rumen protected fat (RPF) in 18 crossbred lactating Karan Fries (Holstein Friesian × Sahiwal) cows. Group 1 was supplemented with 2.5% mustard oil, group 2 with 6% RBLP and group 3 with 3% RPF in the concentrate mixture. The cost of feeding lactating cows in terms of per kg milk and 4% FCM was 10.24, 8.83, 9.67 and 9.40, 7.88, 8.72 in groups 1, 2 and 3, respectively. Cows fed RBLP were more economical by 15.97 and 9.51% to produce whole milk and 19.30 and 10.70% to produce FCM in comparison to groups 1 and 3 cows, respectively.

Yadav *et al.* (2015) evaulated effect of prilled fat supplementation on certain plasma metabolites and milk production in Karan Fries cows. The feeding of prilled fat was economical resulting in an additional income generation of Rs. 94.46/cow/day.

2.14 Protected Protein

Appropriate technological methods such as physical, chemical or combination of both, for the proteinous feeds and their by-products can be employed before their inclusion in the rations of livestock for improving productivity. Highly degradable proteinous oil cakes when ingested by ruminants result in large scale ammonia production much of it gets wasted as urea excreted through urine. Even the animal has to spent energy to convert ammonia into urea in liver. In order to increase the efficiency of protein utilization from the highly degradable cakes, these proteins need to be protected from excessive ruminal degradation and can be used as bypass protein, so that the amino acids from these protein feeds are absorbed intact in the intestines of the animal for tissue protein synthesis as well as for the process of gluconeogenesis in liver. In rumen, the ingested proteins are extensively degraded by microbial proteases to polypeptides, peptides and amino acids which are again deaminated to liberate ammonia and volatile fatty acids.

2.15 Methods of Protein Protection

2.15.1 Naturally protected proteins

The protein of most of the feed resources is bypass to some extent, but some bypass more than others. The natural protection depends upon the following properties of feeds.

1. Surface area available for microbial attack
2. Chemical natures of proteins
3. Physical consistencies of proteins
4. Presence of other dietary components
5. Passage rate from rumen

2.15.1.1 Sources of protected protein

2.15.1.1.1 Lower protein degradability: Maize gluten meal, cottonseed cake, fish meal, coconut cake and maize grain.

2.15.1.1.2 Medium protein degradability: Linseed cake, deoiled rice bran and soybean meal.

2.15.1.1.3 Highly protein degradability: Mustard cake (MC) and Groundnut cake (GNC).

However, those feeds contain 16-53% of total N in the form of acid detergent insoluble nitrogen. This is because of the presence of tannins, particularly the condensed tannins which bind the proteins irreversibly and if fed to animals, are capable of corroding (destroying) the epithelial lining of the gastrointestinal tract. So, tree forages could be used as a source of protected protein only after devising a method for their tannin detoxification, using either some chemical, biological or biotechnological approach. While the proteins of lower protein degradability do not need any protection, highly degradable cakes like MC, GNC and sunflower seed cake need protection against attack of ruminal proteolytic enzymes, for improving their utilization by ruminants.

2.15.2 Heat treatment

The drying of forage is known to increase the protection of the proteins. During the process of manufacturing oil seed meals, they are subjected to different degree of heating which partly explains differences in the degree of protection. Thorough heating of protection against microbial fermentation in the rumen. Heat treatment at 125- 150^{0}C for 2-4 hours could protect proteins very efficiently. High pressure steam treatment with extrusion has shown promising results. The problem with 'heat treatment' is that it may not be cost effective and moreover, it can also over-protect the protein (Sengar and Mudgal, 1982).

Walli and Sirohi (2004) observed that the roasting of soybean at 130°C for 30 min protected its protein from ruminal degradation. Walli (2005) fine-tuned the heat treatment of GN-cake and soybean-cake and found 150°C for 2hr as the optimum temperature-time combination.

2.15.3 Formaldehyde treatment

It is most widely used chemical treatment for the protection of protein. Normally we add 3-4 kg of commercial formalin (37-40% HCHO) used per 100 kg of CP or 1-1.2 g HCHO/ 100g CP. The most successful procedure is developed by Ferguson in 1967. HCHO binding to the proteins by formation of methylene bridge which make them resistant to microbial attach. Generally there is increased fecal nitrogen and decrised urinary nitrogen which indicate effectiveness of protection. The use of formaldehyde to protect dietary protein for rumin ants is based on the premise that bound formaldehyde markedly reduces the solubility of the protein at pH 6.0, there by rendering it highly resistant to microbial attack in the rumen, without significantly reducing its digestibility in the small intestine. Other aldehydes like, acetaldehyde, glutaraldehyde, glyoxal are also effective but they don't possess any advantage over formaldehyde which is comparatively cheaper and easily available. It is based on the principle of Maillard reaction that the aldehyde group reacts with free amino groups and N-terminal groups and the resultant complex formed thus reduces the solubility of the proteins. The formaldehyde protein complex is fairly stable at rumen pH. However, the complex decomposes easily at the acid pH of the abomasum, enabling the protein to be easily digested in the small intestine. Formaldehyde treatment of cakes groundnut cake effectively reduces their protein degradability and feeding of such treated cakes improved performance of animals in terms of growth and milk production. The treated cake has no effect on the health of animals or the milk quality. The method is also low cost and feasible; the technology has gone commercial in India.

Formaldehyde treatment has been used by several workers in India to reduce the protein degradability of high degradable cakes and also to studied the impact of its feeding on the productive performance of dairy animals (Sahoo *et al.*, 2006; Sahoo and Walli, 2007; Shelke *et al.*, 2012a; Kumar *et al.* 2015).

2.16 Effect of Protected Protein on Dry Matter Intake

Kalbande and Thomas (1999) reported that feed intake of animals was increased significantly (P<O.O1) with increase in UDP levels in the concentrate mixtures. Dry matter intake calculated for the entire period of experiment, being 3.70, 3.38 and 3.14 per cent with varying rumen degradable protein (RDP) to undegradable dietary protein (UDP) ratios of 37:63; 52:48 and 70:30 groups.

Wankhede and Kalbande (2001) studied effect of feeding bypass protein with urea treated grass on the performance of Red Kandharri calves. The dry matter intake per day and per 100 kg body weight was 1.94 and 2.22 and 1.90 and 2.35 kg, respectively in control and bypass protein supplemented group, which was significantly higher in bypass protein group.

Kumar *et al.* (2005) conducted an experiment in order to assess the effect of level of RDP:UDP ratio and level of feeding concentrate on nutrient utilization in lactating crossbred cattle. Twenty four medium producing (~10 kg/d, 45 to 135 days postpartum) lactating crossbred cows were divided into four groups of six animals each in a 2×2 factorial completely randomized design. The cows in group 1 were fed concentrate mixture I containing 59:41 RDP:UDP ratio (low UDP) at normal plane (LUDP+NP), in group 2 were fed low UDP ration at 115% of NRC (1989) requirements (LUDP+HP), whereas cows in group 3 were fed concentrate mixture II containing 52:48 RDP:UDP ratio (high UDP) at normal plane (HUDP+NP) and in group 4 were fed high UDP ration at 115% of NRC (1989) requirements (HUDP+HP). Green jowar was fed ad libitum as the sole roughage to all the animals. The experimental feeding trial lasted for 105 days. The total DM intake, roughage DM intake, concentrate DM intake, roughage: concentrate ratio, DM intake/100 kg body weight and DM intake/kg $W^{0.75}$ did not differ significantly among the different treatment groups.

Mishra *et al.* (2006) conducted an on-farm trial for a period of 90 days on 14 crossbred lactating cows to study the effect of bypass protein supplementation on a paddy straw ration on nutrient utilization, milk production and reproduction performances. The daily dry matter intake (kg/100kg BW or g/kgW0.75) of lactating cows did not differ significantly (P<0.05) between the treatment groups and control which clearly indicates that both the concentrate mixtures equally influenced the voluntary feed intake and palatability. The DMI (kg/100 kg BW) of crossbred lactating cows varied in the range of 2.45 to 3.70, further the daily average DMI through roughage and concentrate over the whole experimental period did not show any significant difference between the two groups.

Torane *et al.* (2006) conducted experiment to assess the effect of different levels and sources of bypass protein either with 4% urea treated or untreated wheat straw on performance of 18 crossbred (Holstein Friesian x Deoni) calves

(12.44±0.48 m age;116.88±3,07 kg). One balanced concentrate mixture (CM) with RDP to UDP ratio of 65:35 was compounded and fed to the calves in control group C with untreated wheat straw ad libitum. Calves in experimental groups T_1 and T_2 were fed ad libitum urea treated wheat straw with decorticated groundnut cake and un-decorticated cottonseed cake having RDP to UDP ratio of 78:22 and 52:48, respectively. Voluntary feed intake in groups T_2 and C was statistically similar (4.90 and 4.19 kg per day, respectively) but was significantly ($P<0.01$) higher as compared to T_1(3.07 kg).

Yadav *et al.* (2006) studied effect of feeding bypass protein with urea treated sorghum straw on performance of crossbred (Jersey + Red Kandhari) calves. Twelve crossbred (Jersey x Red Kandhari) calves were used in an experiment to assess the effect of feeding concentrate mixtures varying in RDP to UDP levels either with 4% urea treated or untreated sorghum straw for a period of 90 days. Two isocaloric and isonitrogenous balanced concentrate mixtures (CM-I and CM-II) varying in RDP to UDP ratio viz., 65:35 and 55:45 were compounded. The calves in control group C were fed CM-I with *ad libitum* untreated sorghum straw while those in experimental group T were fed CM-II with *ad libitum* urea treated sorghum straw. Daily dry matter consumption (2.79 vs 3.60 kg) was significantly ($P<0.01$) higher in bypass protein group as compared to control group.

Arewad *et al.* (2011) studied the effect of feeding bypass protein based total mixed ration (TMR) having roughage to concentrate ratio of 50:50 on the growth and nutrient utilization of crossbred calves. Sixteen calves of similar body weight and age were randomly divided into two groups and fed on two dietary treatments for a period of 120 days. Control group (T_1) was fed on TMR having concentrate mixture without bypass protein source, whereas treatment group (T_2) was fed on similar TMR, but with the protein supplements (guar, rapeseed and groundnut) treated with formaldehyde (1% of CP content). The average DMI was 6.48 and 6.21 kg/day and DMI/ 100kg BW was 4.20±0.12 and 3.63±0.14 in groups T_1 and T_2, respectively and variation between groups was non significant.

Sirohi *et al.* (2013) conducted an experiment to study the supplementation effect of formaldehyde treated mustard cake on its chemical composition, nitrogen fractions, in situ degradability, nutrient utilization and milk production performance in crossbred (Karan- Fries) dairy cows. Twelve lactating crossbred cattle were randomly divided into two groups on the basis of milk yield (10 to 12 kg/day) and days of calving (90 days) in a randomized block design. They were fed green berseem fodder, wheat straw and concentrate mixture supplemented with 0.9 kg untreated (raw) mustard cake as fed basis (8 per cent of total diet) in the control group, whereas untreated mustard cake was replaced by formaldehyde treated mustard cake in the treatment group to make both diets isonitrogenous and isocaloric. The roughage to concentrate ratio was 55:45 in both the diets. The total dry matter intake apparently higher in the treatment group but values remained statistically similar in both groups.

Movaliya *et al.* (2013) conducted feeding trial in growing Jaffrabadi heifers at Cattle Breeding Farm to study the effect of feeding cotton seed cake (CSC) and supplementation of bypass methionine and lysine on growth rate, body measurements, nutrients intake, nutrient digestibility, efficiency for feed and nutrient utilization and cost of feeding. Eighteen Jaffrabadi heifers were divided into three equal groups: control (T-I), CSC (T-II) and bypass methionine-lysine supplementation (T-III). The total DM intake was significantly ($P<0.05$) higher in T_3 as compared to that in T_1 and T_2. The DMI per kg metabolic body size was significantly ($P<0.05$) higher in T_3 as compared to that in T_1 and T_2 groups.

Amrutkar *et al.* (2014) investigated the effect of supplementing rumen protected methionine (RPM) and lysine (RPL) on milk production, composition and nutrient utilization in crossbred cows (*Bos taurus* × *Bos indicus*). Eighteen crossbred cows were selected and divided into two groups (9 each) on the basis of most probable production ability (MPPA) and lactation number. Animals in control group (G-1; MPPA 4119 kg) were fed chopped wheat straw, chaffed green maize fodder and concentrate mixture as per requirements (NRC, 2001). However, animals in supplemented group (G-2; MPPA 4120 kg) were fed same ration as control group plus 7 g RPM and 60 g RPL. The experimental period started from 30 days before expected date of parturition to 120 days post parturition. The mean DMI was 13.01 kg and 13.36 kg/d in group G-1 and G-2, respectively, and higher DMI ($P<0.05$) was recorded in group G-2. Average CPI was 1.91 and 1.95 kg/d in the group G-1 and group G-2, respectively, which was higher ($P< 0.01$) in group G-2 than that of group G-1. Sai *et al.* (2014) studied the effect of supplementation of rumen protected methionine plus lysine on nutrient intake in calves. There was no difference in average DM, CP and TDN intake.

2.17 Effect of Protected Protein on Digestibility of Nutrients

Wankhede and Kalbande (2001) studied effect of feeding bypass protein with urea treated grass on the performance of Red Kandhari calves. The digestibility coefficient of DM (59.60 vs 68.78), CP (53.00 vs 59.83) EE (65.08 vs 71.41) and CF (45.82 vs 49.93) was significantly ($P<0.01$) higher in bypass protein supplementing group.

Kumar *et al.* (2005) conducted an experiment in order to assess the effect of level of RDP:UDP ratio and level of feeding concentrate on nutrient utilization in lactating crossbred cattle. Twenty four medium producing (~10 kg/d, 45 to

135 days postpartum) lactating crossbred cows were divided into four groups of six animals each in a 2×2 factorial completely randomized design. The cows in group 1 were fed concentrate mixture I containing 59:41 RDP:UDP ratio (low UDP) at normal plane (LUDP+NP), in group 2 were fed low UDP ration at 115% of NRC (1989) requirements (LUDP+HP), whereas cows in group 3 were fed concentrate mixture II containing 52:48 RDP:UDP ratio (high UDP) at normal plane (HUDP+NP) and in group 4 were fed high UDP ration at 115% of NRC (1989) requirements (HUDP+HP). Green jowar was fed *ad libitum* as the sole roughage to all the animals. The experimental feeding trial lasted for 105 days. They suggest that, increasing UDP level from 41% to 48% of CP in concentrate mixture and also increasing plane of feeding from normal (100%) to 115% of NRC requirements maintain a consistently higher milk production. The digestibilities of DM, OM, CP, CF, EE and NFE and intakes of TDN and DCP did not differ significantly among the different groups and also due to both UDP level and plane of nutrition and also due to their interaction.

Mishra *et al.* (2006) conducted an on-farm trial for a period of 90 days on 14 crossbred lactating cows to study the effect of bypass protein supplementation with a paddy straw ration on nutrient utilization, milk production and reproduction performances. The experimental cows of both the groups were fed paddy straw, mixed grasses and concentrate mixture as per NRC (2001). The concentrate mixture of group 1 animals contained 25 parts of untreated groundnut cake (GNC) whereas that of group 2 contained 25 parts of formaldehyde treated GNC. The digestibility of DM and EE was significantly ($P<0.05$) higher in cows of group 2 than those of group 1. The digestibility of all other nutrients was similar in both the groups.

Torane *et al.* (2006) conducted experiment to assess the effect of different levels and sources of bypass protein either with 4% urea treated or untreated wheat straw on performance of 18 crossbred (Holstein Friesian x Deoni) calves (12.44±0.48 m age;116.88±3,07 kg). One balanced concentrate mixture (CM) with RDP to UDP ratio of 65:35 was compounded and fed to the calves in control group C with untreated wheat straw ad libitum. Calves in experimental groups T1 and T2 were fed *ad libitum* urea treated wheat straw with decorticated groundnut cake and un-decorticated cottonseed cake having RDP to UDP ratio of 78:22 and 52:48, respectively. The digestibility coefficients of DM, OM and EE were significantly ($P<0.01$) higher for ration T_1 and T_2 as compared to ration C. Digestibility of CP was similar for ration T_1 and T_2 but the values were significantly ($P<0.01$) higher than that of ration C. Digestibility of crude fiber ration T_2 was significantly ($P<0.05$) higher as compared to T_1 and C.

Yadav *et al.* (2006) studied effect of feeding bypass protein with urea treated sorghum straw on performance of crossbred (Jersey + Red Kandhari) calves.

Twelve crossbred (Jersey x Red Kandhari) calves (19.1±0.86 months of age and 88.4±1.03 kg body weight) were used in an experiment to assess the effect of feeding concentrate mixtures varying in RDP to UDP levels either with 4% urea treated or untreated sorghum straw for a period of 90 days. Two isocaloric and isonitrogenous balanced concentrate mixtures (CM-I and CM-II) varying in RDP to UDP ratio viz., 65:35 and 55:45 were compounded. The calves in control group C were fed CM-I with *ad libitum* untreated sorghum straw while those in experimental group T were fed CM-II with *ad libitum* urea treated sorghum straw. The digestibility coefficients of CP, EE and CF were significantly higher for bypass protein group than control group.

Yadav and Chaudhary (2010) studied effect of formaldehyde treated groundnut cake on dry matter intake, digestibility of nutrients in crossbred heifers.Eighteen crossbred heifers were divided into 3 groups of 6 heifers each as uniformly as possible with regard to their age and body weight and maintained on 3 respective isonitrogenous and isocaloric rations. Each animal in all the groups were fed standard ration, comprising 5 kg green berseem (*Trifolium alexandrinum*) and wheat *(Triticum aestivum)* straw *ad lib* and in treatment T_1 heifers were given untreated GNC, while in treatment T_2 and T_3 formaldehyde (FA) treated @ 0.5 g FA/100 g CP and FA treated @ 1.0 g FA/100 g CP, respectively, as a source of protein in the concentrate mixtures as per their requirements. The differences in digestibility coefficient for dry matter (DM), crude protein (CP), crude fibre (CF) and nitrogen free extract (NFE) except ether extract (EE) were non significant among different groups. However, ether extract (EE) digestibility was significantly ($P<0.05$) higher in T_3 than T_1 group, but at par for T_3 and T_2 and T_1 and T_2.

Arewad *et al.* (2011) studied the effect of feeding bypass protein based total mixed ration (TMR) having roughage to concentrate ratio of 50:50 on the growth and nutrient utilization of crossbred calves. Sixteen calves of similar body weight and age were randomly divided into two groups and fed on two dietary treatments for a period of 120 days. Control group (T_1) was fed on TMR having concentrate mixture without bypass protein source, whereas treatment group (T_2) was fed on similar TMR, but with the protein supplements (guar, rapeseed and groundnut) treated with formaldehyde (1% of CP content). Nutrient digestibility of DM, CP, CF, EE and NFE were similar in both groups.

Patel *et al.* (2012) conducted on-farm trial on 24 growing buffalo heifers for a period of 24 weeks in a tribal belt of Panchmahal district, Gujarat to study the effect of compounded concentrate mixture with bypass protein (formaldehyde treated) on nutrient utilization and economics of feeding. The selected animals with an average age of 246±4.8 days and 75.2±2.42 kg body weight were divided in to two groups of 12 each. The selected animals were fed with compound concentrate mixture (20% CP), with bypass protein source in treatment group (T_2) and without bypass protein in control group (T_1). Paddy straw was offered ad lib. along with 2.0 kg of green pasture grass in both groups. The digestibility of nutrients like DM, CP, EE and CF was significantly improved (P<0.05) in bypass protein group (T_2).

Sirohi *et al.* (2013) conducted an experiment effect of formaldehyde treated mustard cake on nutrient utilization in crossbred (Karan- Fries) dairy cows. The digestibility of dry matter, organic matter, crude protein, ether extract and neutral detergent fiber was not significantly different in both groups.

Movaliya *et al.* (2013) conducted feeding trial in growing Jaffrabadi heifers at Cattle Breeding Farm feeding cotton seed cake (CSC) and supplementation of bypass methionine and lysine on growth rate, body measurements, nutrients intake, nutrient digestibility, efficiency for feed and nutrient utilization and cost of feeding. Eighteen Jaffrabadi heifers were divided into three equal groups: control (T-I), CSC (T-II) and bypass methionine-lysine supplementation (T-III). The differences in digestibility coefficient for DM, OM, EE, CF, CP, NFE, NDF and ADF differed significantly (P<0.05) between T_1 and T_3, values in T_2 and T_3were similar.

Sai *et al.* (2014) studied the effect of supplementation of rumen protected methionine plus lysine on nutrient utilization in calves. There was no effect on digestibility of nutrients except that of CP which was higher (P<0.05) in treatment group.

2.18 Effect of Protected Protein on Nutrient Intake

Wankhede and Kalbande (2001) studied effect of feeding bypass protein with urea treated grass on the performance of Red Kandhari calves. The nutritive value of DCP and TDN was 7.32 and 47.34% and 9.39 and 52.40% in control and bypass supplementing group which was significantly (P<0.01) higher than control.

Kumar *et al.* (2005) conducted an experiment in order to assess the effect of level of RDP:UDP ratio and level of feeding concentrate on nutrient utilization in lactating crossbred cattle. Twenty four medium producing (~10 kg/d, 45 to 135 days postpartum) lactating crossbred cows were divided into four groups of six animals each in a 2×2 factorial completely randomized design. The cows in group 1 were fed concentrate mixture I containing 59:41 RDP:UDP ratio (low UDP) at normal plane (LUDP+NP), in group 2 were fed low UDP ration at 115% of NRC (1989) requirements (LUDP+HP), whereas cows in group 3 were fed concentrate mixture II containing 52:48 RDP:UDP ratio (high UDP) at normal plane (HUDP+NP) and in group 4 were fed high UDP ration at 115% of NRC (1989) requirements (HUDP+HP). The intakes of DCP and TDN did not differ significantly among the different groups and also due to both UDP level and plane of nutrition and also due to their interaction.

Yadav and Chaudhary (2010) studied effect of formaldehyde treated groundnut cake on dry matter intake, digestibility of nutrients in crossbred heifers. Eighteen crossbred heifers were divided into 3 groups of 6 heifers each as uniformly as possible with regard to their age and body weight and maintained on 3 respective isonitrogenous and isocaloric rations. Each animal in all the groups were fed standard ration, comprising 5 kg green berseem (*Trifolium alexandrinum*) and wheat *(Triticum aestivum)* straw *ad lib.* and in treatment T_1 heifers were given untreated GNC, while in treatment T_2 and T_3 formaldehyde (FA) treated @ 0.5 g FA/100 g CP and FA treated @ 1.0 g FA/100 g CP, respectively, as a source of protein in the concentrate mixtures as per their requirements. The crude protein (CP) intake per 100 kg body weight was significantly ($P<0.05$) higher in T_1 as compared to T_3 group. The DCP and TDN intake (g/d and % B. Wt.) of heifers during experimental period was not influenced by different levels of protected protein.

Amrutkar *et al.* (2014) investigated the effect of supplementing rumen protected methionine (RPM) and lysine (RPL) on milk production, composition and nutrient utilization in crossbred cows (*Bos taurus* × *Bos indicus*). Eighteen crossbred cows were selected and divided into two groups (9 each) on the basis of most probable production ability (MPPA) and lactation number. Animals in control group (G-1; MPPA 4119 kg) were fed chopped wheat straw, chaffed green maize fodder and concentrate mixture as per requirements (NRC, 2001). However, animals in supplemented group (G-2; MPPA 4120 kg) were fed same ration as control group plus 7 g RPM and 60 g RPL. The experimental period started from 30 days before expected date of parturition to 120 days post parturition. Average CPI was 1.91 and 1.95 kg/d in the group G-1 and group G-2, respectively, which was higher ($P< 0.01$) in group G-2 than that of group G-1. Average TDNI was 8.76 kg and 9.05 kg/d in group G-1 and G-2, respectively. The average TDNI was higher ($P<0.01$) by 3.31% G-2 over that of group G-1.

Sai *et al.* (2014) studied the effect of supplementation of rumen protected methionine plus lysine on nutrient intake in calves. There was no difference in average CP and TDN intake.

2.19 Effect of Protected Protein on Body Weight Change and Body Condition Score

Kalbande and Thomas (1999) studied effect of bypass protein on yield and composition of milk in crossbred cows. Jersey crossbred cows (18) were divided into 3 groups (1, 2 and 3) of 6 animals each as uniformly as possible with regard to their body weights and milk production and maintained on 3 respective isonitrogenous and isocalaric concentrate mixtures A, B and C with varying rumen degradable pratein (RDP) to undegradable dietary protein (UDP) ratios of 37:63; 52:48 and 70:30 for 100 days starting fifth day postpartum. The data of fortnightly body weights of animals revealed that, animals maintained on concentrate mixture A with UDP level of 63.38% had a total gain of 10.38 kg (P<O.O1) over a period of 100 days, those fed concentrate mixture B with UDP level of 47.55% and C with UDP level of 29.75%, lost their body weights to the extent of 3.33 and 5.83 kg, respectively, over the same period of experiment.

Sai *et al.* (2014) studied the effect of supplementation of rumen protected methionine plus lysine on growth performance, nutrient utilization and blood metabolites in calves. Animals were offered a basal diet consisting of wheat straw, green fodder and concentrate mixture to meet their nutrient requirements in control group for a period of 90 days. Animals in treatment group were fed the basal diet plus 2 g RPM (net 0.78 g RPM deliverable) and 17 g RPL (net 3.7 g RPL deliverable) per calf per day. In the middle of the experiment, a metabolic trial was conducted. Average daily gain was higher (P<0.05) by 16.1% in treatment group over that of control group.

2.20 Effect of Protected Protein on Milk Yield

Morgan (1985) reported positive response on milk production performance as a result of feeding formaldehyde treated proteins.

Kunju *et al.* (1992) conducted feeding trial with formulated bypass protein feed on straw based ration by using lactating cross bred cows at the stage of 4th month lactation. Bypass protein feed was fed at 5 different level. They reported that milk production and 4% ECM was observed increasing in accordance with the level of bypass protein feed intake. However, the maximum response was noticed in cows that were fed 3 kg bypass protein feed.

Sampath *et al.* (1997) reported significantly higher FCM yield in lactating crossbred cows fed formaldehyde treated GNC (7.8 vs. 9.4 kg/d).

Kalbande and Thomas (1999) studied effect of bypass protein on yield and composition of milk in crossbred cows maintained on 3 respective isonitrogenous and isocalaric concentrate mixtures A, Band C with varying rumen degradable protein (RDP) to undegradable dietary protein (UDP) ratios of 37:63; 52:48 and 70:30 for 100 days starting fifth day postpartum. Result on lactation studies revealed significant (P<O.O1) differences in total milk yield between the groups fed on the 3 concentrate mixtures. A, B and C the average daily milk yield

calculated over a 100 day lactation period being 10.11,7.18, and 6.32 kg, respectively. Efficiency of milk production, though not significant, was also higher with higher levels of UDP over the entire period of experiment. Overall evaluation of results on milk production and feed conversion efficiency indicated that higher levels of UDP in concentrate mixture stimulated feed consumption, resulting in better availability of nutrients for higher milk production. Results also emphasize the need for formulating suitable feed combinations based on their UDP/by-pass protein levels in addition to total protein for optimum milk production and milk composition.

Garg (1998) reviewed that bypass protein feeding protein shown improvement in milk production in medium to high yielding cows.

Gulati *et al.* (2002) overviewed on rumen protected or bypass protein and their potential to increased milk production in India. They observed that feeding optimally protected RUP supplement (1 and 1.5 kg per cow per day) significantly increased milk yield.

Garg *et al.* (2003a) conducted feeding trial using bypass protein supplement on 16 lactating buffaloes which divided into two group based on sunflower meal. The daily average milk yield was 8.5 ± 0.15 and 9.3 ± 0.14 kg per control and protected protein group which was significantly higher in protected protein group.

Garg *et al.* (2004) reported that milk yield was significantly increased by feeding protected sunflower seed meal (Formaldehyde treated) in lactating cows.

Yadav and Chaudhary (2004) reported significant increased milk yield and FCM yield in medium producing cows on feeding formaldehyde treated GNC.

Walli and Sirohi (2004) reported 15% increase in milk yield on feeding of formaldehyde protected mustard cake to crossbred cows.

Garg *et al.* (2005) studied effect of rumen protected protein on milk production in low yielding 16 crossbred cows. Cows yielding 4–5 kg milk per animal per day were divided into two groups of eight each, based on milk yield, fat percentage and stage of lactation. The animals in both the groups were fed basal diet, comprising of 9.0 kg green maize fodder and 5.0 kg paddy straw. In addition to the basal ration, animals in the groups were fed 1.0 kg each of either untreated (Control) or formaldehyde treated (Experimented) rapeseed meal (Brassica campestris). Average increase in milk yield and FCM in experimental group, over control, was 0.70 kg and 1.0 per cent, respectively. Increase in milk yield in experimental group compared to control group was significantly ($P<0.05$) higher. However, no significant effect was observed on the level of FCM.

Sahoo and Walli (2005) reported that milk yield in mustard cake fed goats increased significantly from 1306 g/d in control group to 1439 g/d in formaldehyde treated mustard cake fed group.

Sampath *et al.* (2005) conducted on-farm lactation trials of 6 months duration in two villages (Anagalapura and Menesi of Doddaballapur taluk of Bangalore District in Kamataka State), to study the effect of incorporating a bypass protein ingredient in the ration of crossbred cows on milk production. In both the villages, the cows in control group were fed finger millet straw, maize fodder and mixed local grass supplemented with groundnut cake (GNC) and wheat bran as practiced by the farmers. In experimental group 50% of GNC was replaced with cotton seed extraction (CSE). The milk yield and SNF in experimental group was higher ($P<0.05$) than in control group in Anagalapura (9.3 vs 7.6L/d). However, in Menesi village the differences (10.5 vs 9.3L/d) were statistically non significant, though the fat and total solids in milk were higher ($P<0.05$) in experimental group.

Kumar *et al.* (2005) conducted an experiment in order to assess the effect of level of RDP:UDP ratio and level of feeding concentrate on milk yield in lactating crossbred cattle. The cows in group 1 were fed concentrate mixture I containing 59:41 RDP:UDP ratio (low UDP) at normal plane (LUDP+NP), in group 2 were fed low UDP ration at 115% of NRC (1989) requirements (LUDP+HP), whereas cows in group 3 were fed concentrate mixture II containing 52:48 RDP:UDP ratio (high UDP) at normal plane (HUDP+NP) and in group 4 were fed high UDP ration at 115% of NRC (1989) requirements (HUDP+HP). Green jowar was fed *ad libitum* as the sole roughage to all the animals. The experimental feeding trial lasted for 105 days. However, increase in milk yield with increased UDP level and also with increased plane of nutrition was observed consistently throughout the experimental period. However, there was no significant difference in milk yield among different groups of cows. The overall daily average milk yield in cows fed with low and high UDP diets were 7.91 and 8.99 kg, respectively and at normal and higher plane of feeding the milk yield as 8.15 and 8.75 kg/day, respectively. Thus, there was 13.65% increase in milk yield due to high UDP level and 7.36% due to higher plane of feeding. The daily 4%FCM yield as 9.20 kg for low UDP diet and 10.28 kg for high UDP diet, whereas it was 9.11 kg at normal plane of feeding and 10.37 kg at higher plane of feeding.

Mishra *et al.* (2006) studied effect of bypass protein supplementation on milk production in crossbred cows on paddy straw based ration. The average milk yield (L/d) was significantly higher ($P<0.01$) in group 2 (9.84 ± 0.01) compared to group 1 (8.72 ± 0.01).

Trinacly *et al.* (2006) studied the inûuence of rumen-protected protein supplemented with the amino acids lysine (Lys), methionine (Met) and histidine

(His) added either as a powder or in the form of rumen-protected tablets to the rumen of dairy cows on the yield, composition and technological suitability of milk. The experiment was carried out on three lactating Holstein cows with an average weight of 523 kg ûtted with ruminal and duodenal cannulas. The experiment was divided into 4 periods of 14 d (10 d preliminary period and 4 d experimental period). In the ûrst period one cow received the tablets (T) and the other two received the non-tableted mixture (C) with the same composition. In the subsequent period the design was opposite. Cows were fed a diet based on maize silage, lucerne hay and a supplemental mixture. Powder or tablets consisted of puriûed soya-protein HP 300, Lys, Met and His. Average milk yield in C cows was 16.73 kg and was signiûcantly lower than in T cows (17.8 kg; P<0.05).

Garg *et al.* (2007) studied effect of feeding slow ammonia release and protected protein supplement (SARPP) on milk production efficiency and blood urea nitrogen. Animals in group I were fed 1.0 kg untreated rape seed meal containing 50 g untreated urea, animal in group II were fed 1.0 kg treated rapeseed meal containing 50 g untreated urea and animals in group III were fed 1.0 kg treated rapeseed meal containing 50 g treated urea (SARPP) supplements in place of one kg compounded cattle feed. The daily average milk yield in kg was 6.46 ± 0.34, 7.42 ± 0.34 (P<0.05) and 7.70 ± 0.28 (P<0.01) for the group I, II and III, respectively which indicated significantly increased milk yield by feeding SARPP supplement.

Chandrashekhrain *et al.* (2008) conducted on-farm lactation trial of 4 months duration in Anagalpura and Menesi village of Doddaballapura taluka, Bangalore district in Karnataka state to study the effect of feeding bypass protein (UDP) on the milk production performance of crossbred cows. Cows in control group were fed a concentrate mixture (CMC) which contained 37% of CP as UDP (bypass protein), while those in experimental group were fed a concentrate mixture (CME) containing 50% CP as UDP. The animals in both the groups were fed *ad lib* local mixed grass as the major source of roughage. The average milk yield per day in experimental group (8.80±0.59) was significantly higher than those in control group (7.73±0.54).

Garg and Sherasia (2010) reviewed that feeding bypass protein, daily milk yield in experimental group increased by 1.0 to 1.5 liters per day.

Suresh *et al.* (2011) did a meta-analysis using meta calendar to develop database for effect of bypass protein on milk yield in Indian cattle. The UDP intake (g/animal/day) was categorized into 0-100, 101-200, 201-300, 301-400, 401-500 and 501-600. A database on UDP intake (g/animal/day), milk yield (kg) and fat % were developed. The results indicated increase in milk yield as the UDP intake was increased. The amount of 4% FCM was 6.62 kg (SD: 0.43) at 0-100 g UDP intake and progressively increased to 10 kg when the UDP intake was >600 g/animal/day. The milk production response was observed to be quadratic i.e. milk production increased with increasing UDP intake and thereafter milk production showed decreasing trend for unit increase in UDP levels. From the meta-analysis of the data, it could be concluded that feeding of UDP is beneficial in increasing milk yield and the optimum level of UDP required for production of 10 kg 4% FCM among Indian cattle is about 571 g/animal/day.

Dosky *et al.* (2012) evaluated the effect of protected soybean meal (SBM) on total milk yield (TMY) and composition, milk energy and body weight (BW) in lactating Meriz does. Results revealed that protected soybean meal significantly ($P<0.01$) increased TMY (44.20 vs 34.08 kg) as compared to control.

Sherasia *et al.* (2012) conducted feeding trials using slow ammonia release and protected protein supplement (SARPP) on lactating crossbred cows (n=24) and buffaloes (n=24). In addition to basal ration, animals in groups I, II and III were fed 1.0 kg untreated rapeseed meal containing 50 g untreated urea, 1.0 kg treated meal containing 50 g untreated urea and 1.0 kg treated meal containing 50 g treated urea, respectively. The average milk yield (kg) in three groups of cows was 9.67, 10.61 and 11.05, respectively whereas average milk yield (kg) in three groups of buffaloes was 6.46, 7.42 and 7.70, respectively which was significantly higher ($P<0.05$) in SARPP supplement group.

Vahora *et al.* (2012) studied the effect of feeding a concentrate containing formaldehyde-treated protein meals on milk production in dairy buffaloes in two villages of Dahod district of Gujarat. The experiment was carried out for 90 days using 24 lactating buffaloes in their 2nd and 3rd lactation. The animals were divided in two groups, one control group (CON) fed a traditional concentrate and a treatment group (BYPRO) fed the same concentrate but with the protein meals treated with formaldehyde. It is concluded that feeding buffaloes under farm conditions with protein meals treated with formaldehyde led to a 15% increase in yield of 6% FCM compared with feeding untreated protein meals.

Sirohi *et al.* (2013) conducted an experiment with supplementation effect of formaldehyde treated mustard cake on milk production performance in crossbred (Karan- Fries) dairy cows and reported average milk production (10.05 kg/d) and fat corrected milk yield per day (11.03 kg) was significantly ($P<0.05$) higher in treatment group as compared control group (8.99 and 10.18kg/d).

Amrutkar *et al.* (2014) investigated the effect of supplementing rumen protected methionine (RPM) and lysine (RPL) on milk production, composition and nutrient utilization in crossbred cows (*Bos taurus* × *Bos indicus*). Animals in control group (G-1; MPPA 4119 kg) were fed chopped wheat straw, chaffed green maize fodder and concentrate mixture as per requirements (NRC, 2001). However, animals in supplemented group (G-2; MPPA 4120 kg) were fed same ration as control group plus 7 g RPM and 60 g RPL. The experimental period started from 30 days before expected date of parturition to 120 days post parturition. Milk yield during supplementation period was 15.89 kg/d in group G-1 and 17.69 kg/d in G-2, which was 11.33% higher (P<0.01) in group G-2 over that of group G-1. Average daily 4% FCM yield was 16.21 in group G-1 and 18.24 kg in group G-2. Group G-2 had 12.52% higher (P<0.01) FCM yield over that of group G-1. Average daily ECM yield was 16.20 in group G-1 and 18.11 kg in group G-2. Rumen protected methionine and lysine supplemented group had 11.79% higher (P<0.01) ECM yield over that of control group. The improvement (P<0.01) in milk production was observed in rumen protected methionine and lysine supplemented group. The major reason for higher milk production in groups G-2 may be improved amino acids efficiency on supplementation of rumen protected methionine and lysine.

Qaisi Al and Titi (2014) conducted study to evaluate the effect of supplementing rumen-protected methionine to early lactating Shami goats on milk production, composition, fatty acid profile, and growth performance of their kids. Daily milk production and energy corrected milk were not affected by protected methionine supplementation on goats.

2.21 Effect of Protected Protein on Milk Composition

Cant *et al.* (1991) reported that post ruminal infusion of casein into fat-supplemented diet containing 15.9 CP increased milk CP content by 0.08 percentage units. Thus supplementation of RPAA to diets containing added fat increased milk CP content.

Gulati *et al.* (2002) overviewed on rumen protected or bypass protein and their potential to increased milk production in India. They observed that feeding optimally protected RUP supplement (1 and 1.5 kg per cow per day) significantly increased milk protein and fat content.

Chatterjee and walli (2003) observed increased in milk fat and 0.5 per cent increase in milk protein of lactating buffaloes supplemented with formaldehyde treated mustard cake.

Garg *et al.* (2003a) conducted feeding trial using bypass protein supplement on 16 lactating buffaloes which divided into two group based on sunflower meal. The fat and protein per cent was 6.7 ± 0.05 and 71 ± 0.06 and 3.5 ± 0.01 and 3.7 ± 0.02 for control and protected protein group, respectively in which fat per cent was significantly higher in protected protein group and no significant effect was observed on level of protein percent in milk by feeding protected protein.

Gerg *et al.* (2003b) reported significant increase in milk fat per cent and 0.2 per cent increase in milk protein of lactating cow supplemented with 1.0 kg protected rapeseed meal.

Sampath *et al.* (2005) conducted on-farm lactation trials of 6 months duration in two villages (Anagalapura and Menesi of Doddaballapur taluk of Bangalore District in Kamataka State), to study the effect of incorporating a bypass protein ingredient in the ration of crossbred cows on milk production. In both the villages, the cows in control group were fed finger millet straw, maize fodder and mixed local grass supplemented with groundnut cake (GNC) and wheat bran as practiced by the farmers. In experimental group 50% of GNC was replaced with cotton seed extraction (CSE). The SNF in experimental group was higher ($P<0.05$) than in control group in Anagalapura. However, in Menesi village the differences were statistically nonsignificant, though the fat and total solids in milk were higher ($P<0.05$) in experimental group.

Kumar *et al.* (2005) conducted an experiment in order to assess the effect of level of RDP:UDP ratio and level of feeding concentrate on milk composition in lactating crossbred cattle. The 4% FCM yield and also fat yield did not differ significantly among different dietary treatments and also due to UDP level and plane of nutrition, however, 4% FCM yield was increased by 11.74% with high UDP level and 13.83% with higher plane of feeding. The values for total solids, fat, lactose, solids-not-fat and gross energy contents in milk differed significantly ($P<0.05$) among the different groups and were significantly ($P<0.05$) higher in milk of cows fed LUDP+HP diet followed by HUDP+HP diet. Total solids (14.65 and 13.83%), lactose (5.44 and 4.92%), solids-not-fat (9.44 and 8.83%) and gross energy (887 and 838 kcal/kg) of milk decreased significantly ($P<0.05$) with increased UDP level while total solids (13.84 and 14.64), fat (4.84 and 5.36%) and gross energy (832 and 894 kcal/kg) increased significantly ($P<0.05$) with increase in plane of feeding.

Mishra *et al.* (2006) studied effect of bypass protein supplementation on milk composition in crossbred cows on paddy straw based ratio. The weekly average milk fat, milk protein, SNF and total solids percentages were 4.35±0.06 and 5.48±0.09; 3.33±0.01 and 3.44±0.02; 9.44±0.19 and 9.96±0.12; and 14.03±0.24 and 15.62±0.16 in groups 1 and 2, respectively, which were statistically ($P<0.05$) different.

Trinacly *et al.* (2006) studied the inûuence of rumen-protected protein supplemented with the amino acids lysine (Lys), methionine (Met) and histidine (His) added either as a powder or in the form of rumen-protected tablets to the rumen of dairy cows on the yield, composition and technological suitability of milk. The increase in milk protein yield in T cows was followed closely by progressive increases in casein content (2.46 vs 2.68%).

Garg *et al.* (2007) studied effect of feeding slow ammonia release and protected protein supplement (SARPP) on milk production efficiency and blood urea nitrogen. There was significantly increase in fat and protein per cent in SARPP fed group as compared to other groups.

Chandrashekhrain *et al.* (2008) studied the effect of feeding bypass protein (UDP) on the milk production performance of crossbred cows. The milk fat (4.24±0.29 vs 5.08±0.23), SNF (8.51±0.13 vs 8.97±0.13) and total solids (12.77±0.37 vs 14.08±0.33) were significantly higher in experimental animals than in control group animals.

Garg and Sherasia (2010) reviewed that feeding bypass protein meal, milk protein increased by 0.3-0.4 per cent, milk fat increased by 0.2 per cent.

Dosky *et al.* (2012) evaluated the effect of protected soybean meal (SBM) on total milk yield (TMY) and composition, milk energy and body weight (BW) in lactating Meriz does. Results revealed that protected soybean meal significantly ($P<0.01$) increased milk fat % (4.14±0.13 vs 3.32±0.06), and yield (25.45±0.75 vs 16.08±0.41 g/day), milk protein % (4.86±0.05 vs 4.31±0.04), and yield (30.73±1.02 vs 20.99±0.52 g/day), as compared to control.

Sherasia *et al.* (2012) conducted feeding trials using slow ammonia release and protected protein supplement (SARPP) on lactating crossbred cows (n=24) and buffaloes (n=24). Animals were divided into three groups of eight each, based on milk yield, fat percentage and stage of lactation. In addition to basal ration, animals in groups I, II and III were fed 1.0 kg untreated rapeseed meal containing 50 g untreated urea, 1.0 kg treated meal containing 50 g untreated urea and 1.0 kg treated meal containing 50 g treated urea, respectively. The average fat (%) in three groups of cows was 4.30, 4.56 and 4.58, respectively whereas the average fat (%) in three groups of buffaloes was 6.64, 6.81 and 6.86, respectively which was significantly higher ($P<0.05$) in SARPP supplement group

Sirohi *et al.* (2013) studied the effect of supplementation of formaldehyde treated mustard cake on milk composition in crossbred (Karan- Fries) dairy cows and identified that milk composition almost remained unaffected and was similar in both groups.

Amrutkar *et al.* (2014) investigated the effect of supplementing rumen protected methionine (RPM) and lysine (RPL) on milk production, composition and nutrient utilization in crossbred cows (*Bos taurus* × *Bos indicus*). Animals in control group (G-1; MPPA 4119 kg) were fed chopped wheat straw, chaffed green maize fodder and concentrate mixture as per requirements (NRC, 2001). However, animals in supplemented group (G-2; MPPA 4120 kg) were fed same ration as control group plus 7g RPM and 60g RPL. The experimental period started from 30 days before expected date of parturition to 120 days post parturition. The milk fat per cent was higher (P<0.01) by 2.18 per cent in group G-2 than that of group G-1 (4.22 vs 4.13%). However milk protein, milk lactose and SNF (g/100 g) content of milk were similar in both the groups. Mean total solids content in group G-1 and G-2 were 13.03 and 13.14%, respectively which was higher (P<0.01) in group G-2 than that of group G-1.

Qaisi Al and Titi (2014) conducted study to evaluate the effect of supplementing rumen-protected methionine to early lactating Shami goats on milk production, composition, fatty acid profile, and growth performance of their kids. Milk percentages of fat, protein, total solids, and casein were also not affected by protected methionine supplementation. Likewise, fat and protein yields were not different among dietary treatments of protected methionine on goats.

2.22 Effect of Protected Protein on Fatty Acid Profile

Trinacly *et al.* (2006) studied the inûuence of rumen-protected protein supplemented with the amino acids lysine (Lys), methionine (Met) and histidine (His) added either as a powder or in the form of rumen-protected tablets to the rumen of dairy cows on the yield, composition and technological suitability of milk. The content of short- and medium-chain fatty acids was higher and the proportion of C18:1, C18:2, C18:3n3 and C20:1 was lower in the control than in the treatment (P<0.05) animals. The proportion of UFA (both MUFA and PUFA) was increased and that of SFA in milk (P<0.05) was decreased in the treatment animals. The total SFA:UFA ratio was higher (P<0.05) in control compared with treatment cows (1.89 vs 1.52).

Qaisi and Titi (2014) conducted study to evaluate the effect of supplementing rumen-protected methionine to early lactating Shami goats on milk production, composition, fatty acid profile, and growth performance of their kids. No differences in milk fatty acids composition were observed by supplementing rumen protected methionine in the experiment on goats.

2.23 Effect of Protected Protein on Reproduction

Kaur and Arora (1995) reported improvement in conception rate and reduced the number of open days in crossbred cows by feeding bypass protein. Feeding more than the recommended amounts of protein or inclusion of excess degradable protein in the diet adversely affect animal reproduction. More ammonia is released in the rumen through the process of deamination by bacterial enzymes.

Mishra *et al.* (2006) studied effect of bypass protein supplementation on reproduction in crossbred cows fed paddy straw based ration and observed average post- partum oestrus period (d), average days of experimental feeding up to the first oestrus (d), services per conception (no) were 49.71±7.01 and 39.29±4.89; 120.00±0.69 and 109.28±0.61; and 1.8±0.02 and 1.14±0.14 in the groups 1 and 2, respectively. Seven cows in group 2 compared to 5 cows in group 1 were found pregnant registering a conception rate of 100% in the former.

Walli (2008) observed that feeding bypass protein increased the sexual behavior and seminal attributes in mehsana buffalo bulls and in crossbred bucks.

2.24 Effect of Protected Protein on Various Blood Parameters

Garg (1998) reviewed that UDP did not influence blood glucose, blood urea and total protein levels.

Wankhede and Kalbande (2001) studied effect of feeding bypass protein with urea treated grass on the performance of Red Kandhari calves. The blood urea nitrogen levels was significantly ($P<0.01$) lower in calves fed with bypass protein group.

Torane *et al.* (2006) conducted experiment to assess the effect of different levels and sources of bypass protein either with 4% urea treated or untreated wheat straw on performance of 18 crossbred (Holstein Friesian x Deoni) calves. Calves in experimental groups T_1 and T_2 were fed *ad libitum* urea treated wheat straw with decorticated groundnut cake and un-decorticated cottonseed cake having RDP to UDP ratio of 78:22 and 52:48, respectively. The post feeding blood urea nitrogen level in calves of group T_2 was significantly ($P<0.01$) lower as compared to those in control and T_1.

Yadav *et al.* (2006) studied effect of feeding bypass protein with urea treated sorghum straw on performance of crossbred (Jersey + Red Kandhari) calves to assess the effect of feeding concentrate mixtures varying in RDP to UDP levels either with 4% urea treated or untreated sorghum straw for a period of 90 days. Two isocaloric and isonitrogenous balanced concentrate mixtures (CM-I and CM-II) varying in RDP to UDP ratio viz., 65:35 and 55:45 were compounded. The calves in control group C were fed CM-I with *ad libitum* untreated sorghum straw while those in experimental group T were fed CM-II with *ad libitum* urea treated sorghum straw. The blood urea nitrogen concentration (mg/dl) at various post feeding intervals were significantly lower ($P<0.01$) in calves fed bypass protein ration.

Garg *et al.* (2007) studied effect of feeding slow ammonia release and protected protein supplement (SARPP) on milk production efficiency and blood urea nitrogen. Animals in group I were fed 1.0 kg untreated rapeseed meal containing 50 g untreated urea, animal in group II were fed 1.0 kg treated rapeseed meal containing 50 g untreated urea and animals in group III were fed 1.0 kg treated rapeseed meal containing 50 g treated urea (SARPP) supplements in place of one kg compounded cattle feed. On feeding SARPP supplement, BUN (mg/dl) was significantly (P<0.05) low (9.64 ± 0.20) as compared to group I (10.33 ± 0.21) and group II (10.48 ± 0.13) due to SARPP supplement avoided excessive loss of both nitrogen and energy which increase energy and nitrogen balance using increased in milk production.

Sherasia *et al.* (2012) conducted feeding trials using slow ammonia release and protected protein supplement (SARPP) on lactating crossbred cows and buffaloes. In addition to basal ration, animals in groups I, II and III were fed 1.0 kg untreated rapeseed meal containing 50 g untreated urea, 1.0 kg treated meal containing 50 g untreated urea and 1.0 kg treated meal containing 50 g treated urea, respectively. In cows, the level of blood urea-N (mg/dl) was 12.23, 12.29 and 11.63 in groups I, II and III, respectively and in buffaloes, the level of blood urea nitrogen (mg/dl) was 10.33, 10.48 and 9.64 in groups I, II and III, respectively which was significantly lower (P<0.05) in SARPP supplement group.

Movaliya *et al.* (2013) conducted feeding trial in growing Jaffrabadi heifers at Cattle Breeding Farm feeding cotton seed cake (CSC) and supplementation of bypass methionine and lysine supplementation on growth rate, body measurements, nutrients intake, nutrient digestibility, efficiency for feed and nutrient utilization and cost of feeding. No significant difference was observed in biochemical parameters like total protein, albumin, globulin and creatinine except BUN.

Sai *et al.* (2014) studied the effect of supplementation of rumen protected methionine plus lysine on blood metabolites in calves. Plasma triglycerides, total protein, concentration increased (P<0.01) and BUN concentration decreased (P<0.05), whereas no difference in blood glucose, albumin, globulin, NEFA concentration and A: G ratio in rumen protected methionine plus lysine supplemented groups.

2.25 Economics of Protected Protein Feeding

Gulati *et al.* (2002) overviewed on rumen protected or bypass protein and their potential to increased milk production in India. Economic analysis showed that feeding 1 kg of RUP supplement, compared to the naturally protected meal, produced the maximal net daily gain, being Rs. 9.61 per cow per day.

Garg *et al.* (2003a) reported average net daily income increased by Rs. 14.49 on feeding 1.0 kg protected sunflower meal in lactating buffaloes.

Sampath *et al.* (2005) reported with supplementation of bypass protein, the farmer's income increased by Rs. 17.81 and Rs. 15.80/cow/d, respectively, in

Anagalapura and Menesi villages, clearly indicating that scientific intervention i.e. small alteration in the existing feeding practices could improve the performance of animals and farmer's income.

Mishra *et al.* (2006) reported cows earned Rs. 12.75 more profit per cow per day by feeding bypass protein than control group, accruing overall excess profit of Rs. 1147.50 in 90 days, indicating a clear advantage of bypass protein supplementation.

Torane *et al.* (2006) conducted experiment to assess the effect of different levels and sources of bypass protein either with 4% urea treated or untreated wheat straw on performance of crossbred calves. Calves in experimental groups T_1 and T_2 were fed *ad libitum* urea treated wheat straw with decorticated groundnut cake and un-decorticated cottonseed cake having RDP to UDP ratio of 78:22 and 52:48, respectively. The cost of composite ration per kg gain in body weight for calves in groups control, T_1 and T_2 were found to be Rs. 24.85, 17.55 and 15.95, respectively, indicating economic superiority of T_2 over control and T_1.

Chandrashekhrain *et al.* (2008) studied the effect of feeding bypass protein (UDP) on the milk production performance of crossbred cows. The milk yield was increased by 1.07 litres, feed cost was reduced by Rs 2.20 and the overall income of the farmers was increased by Rs 12.40/cow/day in experimental group.

Garg and Sherasia (2010) reviewed that feeding bypass protein meals, daily net income (per animal) increased by Rs. 10-11 in animals yielding 8-10 liters and Rs. 5-6 in animals yielding 4-5 liters.

Arewad *et al.* (2011) reported significantly higher feed cost ($P<0.05$) in group treated than in control on feeding bypass protein based total mixed ration to crossbred calves. The cumulative feed cost (Rs. /head) was significantly ($P<0.05$) higher in treatment group T_2 (Rs.3527.96) than in control group T_1 (Rs. 3073.35), which may be attributed to the extra cost involved in the process of the formaldehyde treatment. The feed cost (Rs/kg gain) was Rs. 61.46 and 57.77 in control and treatment group, respectively.

Patel *et al.* (2012) conducted on-farm trial on 24 growing buffalo heifers for a period of 24 weeks in a tribal belt of Panchmahal district, Gujarat to study the effect of compounded concentrate mixture with bypass protein (formaldehyde treated) on nutrient utilization and economics of feeding. The selected animals were fed with compound concentrate mixture (20% CP), with bypass protein source in treatment group (T_2) and without bypass protein in control group (T_1). Paddy straw was offered *ad lib* along with 2.0 kg of green pasture grass in both groups. The daily feed cost per kg gain was significantly ($P<0.01$) lower in T_2 (Rs.27.79) as compared to T_1 (Rs.31.45).

Movaliya *et al.* (2013) conducted feeding trial in growing Jaffrabadi heifers at Cattle Breeding Farm to study the effects of feeding cotton seed cake (CSC) and supplementation of bypass methionine and lysine supplementation on growth rate, body measurements, nutrients intake, nutrient digestibility, efficiency for feed and nutrient utilization and cost of feeding. The cost of feeding per kg live weight gain was lowest in (CSC) group and the highest in bypass methionine and lysine supplementation group than control.

Amrutkar *et al.* (2014) investigated the effect of supplementing rumen protected methionine (RPM) and lysine (RPL) on milk production, composition and nutrient utilization in crossbred cows (*Bos taurus* × *Bos indicus*). Animals in control group (G-1; MPPA 4119 kg) were fed chopped wheat straw, chaffed green maize fodder and concentrate mixture as per requirements (NRC, 2001). However, animals in supplemented group (G-2; MPPA 4120 kg) were fed same ration as control group plus 7 g RPM and 60 g RPL. The experimental period started from 30 days before expected date of parturition to 120 days post parturition. Net return over feed cost of milk yield per animal per day in groups G-1 and G-2 was Rs. 192.1 and 215.8, respectively, indicating that it was higher by Rs. 23.7 in group G-2 over that of group G-1. The feed cost (Rs.) per kg FCM produced was Rs. 10.94 and Rs. 10.92, respectively in group G-1 and G-2 and cost benefit ratio was also similar (1.08) in both the groups. The net return from supplementation was markedly higher by 12.30% than that of the control ration. The net profit in 120 days from 9 cows on supplementation of rumen protected methionine and lysine was Rs. 25544.6.

2.26 Protected Fat and Protein

2.27 Effect of Protected Fat and Protein on Dry Matter Intake

Shelke *et al.* (2012a) studied effect of supplementing protected nutrients on productive and reproductive performance on 18 Murrah buffaloes. There was no significant effect of protected nutrient supplementation on average DM intake (14.42 ± 0.10 vs 14.70 ± 0.01 kg/d).

Vahora *et al.* (2013) conducted on-farm feeding trial at the Jalat and Borkheda villages of Dahod district in Gujarat for 150 days to know the effect of feeding formaldehyde treated *guar* meal (*Cyamopsis tetragonoloba*) and commercial bypass fat supplement Mehsani buffaloes in their 2nd to 3rd lactation belonging to different farmers. The buffaloes yielding 8–10 kg milk/day were randomly allotted to three dietary treatments (6 in each group) i.e. T_1 (Control) and

T_2 (bypass protein) and T_3 group (bypass protein and bypass fat), based on daily milk yield, fat % and body weight and offered isonitrogenous and isocaloric ration. The cumulative DMI of experimental buffaloes was 1818.79, 1989.19 and 1925.02 kg/head for 150 days in T_1, T_2 and T_3 groups, respectively which differed from each other.

Nam *et al.* (2014) evaluated effect of ruminally protected amino acids (RPAAs) and ruminally protected fat (RPF) on milk yield and milk composition (in vivo) on mid lactating Holstein dairy cows accordingly to mean milk yield and number of postpartum. Supplementation of RPAAs and RPF did not affect ($P>0.05$) the dry matter digestibility.

2.28 Effect of Protected Fat and Protein on Digestibility of Nutrients

Shelke *et al.* (2012) studied effect of supplementing protected nutrients during 60 days pre partum to 90 days postpartum on reproductive performance on Murrah buffaloes. Buffaloes in group C (MPPA 2204.17 kg) were fed wheat straw, green maize fodder and concentrate mixture as per requirements. Buffaloes in group S (MPPA 2210.64 kg) were fed same ration as group C plus 2.5% rumen protected fat (DM intake basis) and formaldehyde treated mustard and groundnut oil cake (1.2 g HCHO/100 g CP). Group S buffaloes were supplemented protected nutrients. The digestibility of dry matter, crude protein, crude fiber were not affected, whereas ether extract digestibility was higher ($P<0.05$) in protected fat and protein group.

Gajera *et al.* (2013) evaluated the incorporation of rumen protected lysine, methionine and fat on nutrient digestibility coefficient of nutrients in Jaffrabadi buffalo heifers. The digestibility of DM, CP, CF, EE and NFE were affected significantly ($P<0.05$) in general by supplementation of bypass methionine, lysine and fat.

Grewal *et al.* (2014) conducted 120 days field study on crossbred lactating cows to evaluate the production economics of supplemental bypass fat and niacin. The control group was fed wheat straw, concentrate mixture, bread waste, biscuit waste, mixed green fodder and baby corn waste while the treatment group received the same ration with additional 200g bypass fat and 12g niacin daily. The digestibility of DM and CP were comparable in both the groups. The digestibility of EE was significantly ($P<0.01$) higher in treatment in comparison to control group (71.84 and 79.25 % respectively).

Nam *et al.* (2014) evaluated effect of ruminally protected amino acids (RPAAs) and ruminally protected fat (RPF) on milk yield and milk composition (in vivo) on mid lactating Holstein dairy cows. Supplementation of RPAAs and RPF did not affect ($P>0.05$) the dry matter digestibility.

2.29 Effect of Protected Fat and Protein on Nutrient Intake

Shelke and Thakur (2011) studied effect of supplementing protected nutrients on nutrients intake on 18 Murrah buffaloes. The CP intake was higher ($P<0.05$) in supplemented group (2.03 kg/d) than that of control (1.75 kg/d). The TDN intake was higher ($P<0.01$) by 16.05% in supplemented group (10.13 kg/d) than that of control (8.72 kg/d).

Gajera *et al.* (2013) evaluated the incorporation of rumen protected lysine, methionine and fat for a period of 105 days on nutrient utilization, nutrient intake and digestibility coefficient of nutrients in Jaffrabadi buffalo heifers. The heifers under control group (T_1) were fed concentrate as per the feeding standards of NRC (1989) to meet out DCP requirement + 15 Kg seasonal green. In addition to the ration of T_1, heifers were fed bypass fat @ 100 g/head/day (T_2), bypass lysine and methionine @ 15 and 5g/head/day, respectively (T_3) and bypass lysine and methionine as per T_3 along with bypass fat as per T_2 (T_4). On an average the animals under T_1, T_2, T_3 and T_4 consumed 1137.96 ± 59.69, 1153.04 ± 73.06, 1154.60 ± 75.26 and 1185.9 ± 55.83 g CP/d, respectively; 691.84 ± 36.56, 755.83 ± 47.89, 873.44 ± 56.43 and 922.01 ± 43.41 g DCP/d, respectively and 5.38 ± 0.24, 5.83 ± 0.28, 5.80 ± 0.34 and 6.36 ± 0.25 kg TDN/d, respectively, however, the differences between the groups in respect of CP intake and TDN intake were reported non significant.

2.30 Effect of Protected Fat and Protein on Body Weight and Body Condition Score

Shelke *et al.* (2012a) reported higher average birth weights of the calves by 10.8% in supplemented group (31.94 vs. 35.38 kg) which was due to higher energy intake in supplemented group during pre partum period on bypass fat supplementation @ 2.5% DMI.

Grewal *et al.* (2014) conducted 120 days field study on crossbred lactating cows to evaluate the production economics of supplemental bypass fat and niacin. The treatment group gained more BW as compared to control (18.47±9.91 vs. 8.67±8.76, respectively) over the 120d period but the difference was non-significant due to higher standard error within the groups.

2.31 Effect of Protected Fat and Protein on Milk Yield

Garg *et al.* (2002) reported that by supplementing bypass nutrients (fats encapsulated in a aldehyde treated protein) @ 1000 g/d to lactating cows and buffaloes increased milk yield significantly.

Strusiñska *et al.* (2006) studied on 37 Holstein-Friesian cows during the first 120 days of lactation. The cows were fed balanced diets composed of haylage and maize silage (together 52% dry mater) and a concentrate containing 19% crude protein (group 1), a concentrate containing 17.5% crude protein and 1 kg of fat-protein supplement Megapro Plus® (group 2) or a concentrate containing 19% crude protein and 1 kg of fat-protein supplement Megapro Plus® (group 3). The mean daily milk yield recorded in groups 2 and 3 amounted to 32.6 kg/d and was by 10.3% higher than in the control group (P d" 0.01). FCM yield increased by 15.5% and 12.1% in groups 2 and 3, respectively, in comparison with group 1.

Shelke *et al.* (2012) studied effect of supplementing protected nutrients on milk performance on 18 Murrah buffaloes. Buffaloes in group C (MPPA 2204.17 kg) were fed wheat straw, green maize fodder and concentrate mixture as per requirements. Buffaloes in group S (MPPA 2210.64 kg) were fed same ration as group C plus 2.5% rumen protected fat (DM intake basis) and formaldehyde treated mustard and groundnut oil cake (1.2 g HCHO/100 g CP). Group S buffaloes were supplemented protected nutrients 60 days pre partum to 90 days postpartum and was monitored up to 210 days of lactation. Milk yield during supplementation period (90 days) in S group was 13.11 kg/d and was 19% higher (P<0.01) than the C group (11.01 kg/d), whereas after supplement withdrawal (120 days), it was 11.04 kg/d and was 15% higher (P<0.01) than the C group (9.61 kg/d).

Gerg *et al.* (2012b) evaluated the effect of supplementing bypass fat, with or without rumen protected choline chloride on milk yield, composition, milk fatty acid profile and metabolic profile in early lactating HF crossbred cows. In addition to basal ration, cows in group II were fed 100 g bypass fat supplement, whereas, cows in group III were fed 100 g bypass fat and 10 g rumen protected choline (RPC) supplement per animal per day. Average increase in daily milk yield in group II and III over a 90 days experimental period were 1.48 kg (P<0.01) and 1.77 kg (P<0.01) as compared to group I.

Dhulipalla *et al.* (2013) conducted experiment to study the effect of feeding Milk Dhara supplement on milk yield and composition in mid lactating buffaloes. The buffaloes in the treatment group were fed same basal diet plus Milk Dhara (M/S Zydus AHL, Ahmedabad) containing rumen stable bypass fat and protein at 100 g/buffalo/day. The average daily milk yield was higher (P<0.05) in the treatment group of buffaloes (9.33 kg) as compared to the control (9.02 kg). The daily average 6% FCM yield increased (P<0.01) in buffaloes fed Milk Dhara supplemented diet compared to those in the control group (9.12 vs. 8.61 kg/d). The average daily milk yield was 3.44% higher in treatment group than that of control group. Average 6% FCM yield was enhanced by 5.92% in treatment group as compared to the control group.

Vahora *et al.* (2013) conducted on-farm feeding trial at the Jalat and Borkheda villages of Dahod district in Gujarat state for 150 days to know the effect of feeding formaldehyde treated *guar* meal (*Cyamopsis tetragonoloba*) and commercial bypass fat supplement on 18 Mehsani buffaloes in their 2nd to 3rd lactation belonging to different farmers. The buffaloes yielding 8–10 kg milk/day were randomly allotted to three dietary treatments (6 in each group) i.e. T_1 (Control) and T_2 (Bypass protein) and T_3group (Bypass protein and bypass fat), based on daily milk yield, fat % and body weight and offered isonitrogenous and isocaloric ration. The daily yield of milk and 6% FCM was significantly higher ($P<0.01$) in T_2 and T_3 as compared to that of T_1, where as that of T_2 and T_3 were similar. This increase in whole milk, 6% FCM and fat yield were consistent during the whole experimental period.

Grewal *et al.* (2014) conducted 120 days field study on crossbred lactating cows to evaluate the production economics of supplemental bypass fat and niacin. The treatment group received the same ration with additional 200g bypass fat and 12g niacin daily. The effect of supplemental bypass fat and niacin on the production performance of crossbred cows indicated higher average milk production ($P<0.05$) in treatment than control group animals (16.84 vs 15.29 kg, respectively). This difference of 1.55 kg milk/d translated into about 186 kg more milk during the 120-d experimental duration due to the supplementation of bypass fat and niacin.

Nam *et al.* (2014) evaluated effect of ruminally protected amino acids (RPAAs) and ruminally protected fat (RPF) on milk yield and milk composition (in vivo) on mid lactating Holstein dairy cows. Milk yield (30.52 vs 33.29 kg) and 4% fat corrected milk (29.73 vs 32.63) were significantly ($P<0.01$), ($P<0.05$) higher in supplemented RPAAs and RPF feeding group as compared to control group.

2.32 Effect of Protected Fat and Protein on Milk Composition

Strusińska *et al.* (2006) studied on 37 Holstein-Friesian cows during the first 120 days of lactation. The cows were fed balanced diets composed of haylage and maize silage (together 52% dry mater) and a concentrate containing 19% crude protein (group 1), a concentrate containing 17.5% crude protein and 1 kg of fat-protein supplement Megapro Plus® (group 2) or a concentrate containing 19% crude protein and 1 kg of fat-protein supplement Megapro Plus® (group 3). Diet supplementation with Megapro Plus® had no significant effect on the levels of fat, lactose, protein, nitrogen fractions and urea, or some physicochemical properties of milk (density, pH, heat stability). A decrease was noted in solids-non-fat (group 2).

Shelke *et al.* (2012) reported no effect on total solid, protein, solid-not fat (SNF) and lactose contents, whereas milk fat and milk fat yield was increased significantly (P<0.05) by feeding rumen protected fat and protein.

Dhulipalla *et al.* (2013) conducted experiment to study the effect of feeding Milk Dhara supplement on milk yield and composition in mid lactating buffaloes. The buffaloes in the treatment group were fed same basal diet plus Milk Dhara (M/S Zydus AHL, Ahmedabad) containing rumen stable bypass fat and protein at 100 g/buffalo/day. The daily average fat content of milk in graded Murrah buffaloes was higher (P<0.01) in Milk Dhara supplemented group (5.78%) when compared to the control group (5.59%). Similarly, % Total solids (P<0.05) and % Protein (P<0.01) content of milk were higher in Milk Dhara supplemented group of buffaloes compared to control. The average SNF % was 10.3 and 10.5% in control and treatment groups, respectively.

Vahora *et al.* (2013) conducted on-farm feeding trial at the Jalat and Borkheda villages of Dahod district in Gujarat state for 150 days to know the effect of feeding formaldehyde treated *guar* meal (*Cyamopsis tetragonoloba*) and commercial bypass fat supplement to recently calved 18 Mehsani buffaloes in their 2nd to 3rd lactation belonging to different farmers. The buffaloes yielding 8–10 kg milk/day were randomly allotted to three dietary treatments (6 in each group) i.e. T_1 (Control) and T_2 (bypass protein) and T_3 group (bypass protein and bypass fat), based on daily milk yield, fat % and body weight and offered isonitrogenous and isocaloric ration. The milk fat % of experimental buffaloes was higher (P<0.01) than that of control. The TS % was higher (P<0.01) in T_3 as compared to control (T_1). The SNF content was not affected by different treatments.

Grewal *at al.* (2014) studied effect of bypass fat and niacin on milk composition of crossbred lactating cows. The treatment group received the same ration with additional 200g bypass fat and 12g niacin daily. The milk composition data revealed that milk fat per cent was significantly (P<0.05) higher (3.31 vs 370) in treatment group and all other components protein (3.30 vs 3.29), lactose (5.10 vs 5.11) and SNF (8.80 vs 8.81) per cent were similar in both the groups.

Nam *et al.* (2014) studied effect of ruminally protected amino acids (RPAAs) and ruminally protected fat (RPF) on milk composition (in vivo) on mid lactating Holstein dairy cows. Milk fat % (3.73 vs 3.87) and milk protein % (3.02 vs 3.05) were non significant among all treatment group but milk fat yield (1.14 vs 1.28 kg) and milk protein yield (0.92 vs 1.02 kg) were significantly higher in supplemented RPAAs and RPF group.

2.33 Effect of Protected Fat and Protein on Fatty Acid Profile

Strusiñska *et al.* (2006) studied on 37 Holstein-Friesian cows during the first 120 days of lactation. The cows were fed balanced diets composed of haylage and maize silage (together 52% dry mater) and a concentrate containing 19% crude protein (group 1), a concentrate containing 17.5% crude protein and 1 kg of fat-protein supplement Megapro Plus® (group 2) or a concentrate containing 19% crude protein and 1 kg of fat-protein supplement Megapro Plus® (group 3). Megapro Plus® supplementation of a diet with a reduced (to 3%) "00" rapeseed meal content in concentrate (group 2) resulted in a significant increase in the concentrations of unsaturated fatty acids (especially C18:1 and C18:2) and hypocholesterolaemic acids (DFA) in milk, recorded on the 120th day of lactation.

Gerg *et al.* (2012) study the effect of bypass fat with rumen protected choiline chloride on milk yield, milk composition and metabolic profile in crossbred cows. They reported significant improvement was observed in poly unsaturated fatty acid (PUFA) content in groups II (100 g bypass fat or @ 34.26%) and III (100 g bypass fat + 10 g RPC or @ 37.45%). Total unsaturated fatty acids increased by about 15.29 and 15.71% in groups II and III, respectively. Long chain fatty acids (LCFA; C16:0 to C20:0) and mono-unsaturated fatty acids (MUFA; C14:1, C16:1 & C18:1) contents were higher in experimental groups as compared to control.

Shelke *et al.* (2012b) studied effect of supplementing protected nutrients on reproductive performance on 18 Murrah buffaloes. Buffaloes in group C (MPPA 2204.17 kg) were fed wheat straw, green maize fodder and concentrate mixture as per requirements. Buffaloes in group S (MPPA 2210.64 kg) were fed same ration as group C plus 2.5% rumen protected fat (DM intake basis) and formaldehyde treated mustard and groundnut oil cake (1.2 g HCHO/100 g CP). Group S buffaloes were supplemented protected nutrients 60 days pre partum to 90 days postpartum and was monitored up to 210 days of lactation. The supplement produced noticeable changes in the fatty acid profile of the milk fat, i.e., reduction in the concentration of saturated fatty acids (SFA) by 19% and an increase in that of unsaturated fatty acids (USFA) by 36%.

2.34 Effect of Protected Fat and Protein on Reproduction

Staples *et al.* (1998) reported that better reproductive performance is associated with higher plasma cholesterol levels, as it acts as a precursor of steroid hormones.

Shelke *et al.* (2012a) studied effect of supplementing protected nutrients on reproductive performance on 18 Murrah buffaloes. Time required for expulsion of fetal membranes was significantly ($P<0.01$) reduced in supplemented group.

Less number of cases of retention of fetal membranes and metritis were observed in supplemented buffaloes. The time required for commencement of cyclicity was reduced (P<0.01) by 8.56 days in supplemented group than that of control group. The commencement of cyclicity is related with the process of involution of uterus, as the duration for uterine involution was reduced on bypass fat supplementation, it may responsible for relatively early commencement of cyclicity. Service period was shorter (P<0.01) by 28.17 days in supplemented group (136.50 days) than that of control group (164.67 days), indicating that lesser time was required for the animals in supplemented group for conception. Buffaloes of supplemented group required less AI per conception (2.48) than those of control group (3.80) indicating better efficiency of AI in supplemented group. The conception rate during the experimental period of 210 days was 55.55 and 77.77% in control group and supplemented group, respectively which indicate that protected nutrient supplementation increased the conception rate in buffaloes.

Grewal *et al.* (2014) conducted 120 days field study on crossbred lactating cows to evaluate the production economics of supplemental bypass fat and niacin. The treatment group received the same ration with additional 200g bypass fat and 12g niacin daily. A total of 14 and 11 number of AI were done in control and treatment groups of which 5 and 6 animals become pregnant in the respective groups up to the termination of experiment. The conception rate calculated was 35.7 and 55.6% in control and treatment groups, respectively.

2.35 Effect of Protected Fat and Protein on Various Blood Parameters

Shelke *et al.* (2012a) studied effect of supplementing protected nutrients on reproductive performance on 18 Murrah buffaloes. Buffaloes in group C (MPPA 2204.17 kg) were fed wheat straw, green maize fodder and concentrate mixture as per requirements. Buffaloes in group S (MPPA 2210.64 kg) were fed same ration as group C plus 2.5% rumen protected fat (DM intake basis) and formaldehyde treated mustard and groundnut oil cake (1.2 g HCHO/100 g CP). Group S buffaloes were supplemented protected nutrients 60 days pre partum to 90 days postpartum and carry over effect of supplementation on milk production and reproductive parameters was monitored up to 210 days of lactation. There was no effect on plasma glucose, non-esterified fatty acids, triglycerides and cholesterol concentrations between two groups, whereas blood urea nitrogen concentration was lower (P<0.01) in S group.

Gerg *et al.* (2012) studied the effect of supplementing bypass fat, with or without rumen protected choline chloride on metabolic profile in early lactating HF crossbred cows. Non-esterified fatty acids (NEFA) in blood serum were reduced by 16.12 and 24.19% ($P<0.01$) in groups of supplementing bypass fat and rumen protected choline chloride, respectively. There was reduction ($P<0.01$) in cholesterol levels in blood serum in animals of groups supplementing bypass fat and rumen protected choline chloride, respectively, as compared to control group. Blood glucose and urea nitrogen were not affected by the dietary treatments.

Grewal *et al.* (2014) conducted 120 days field study on crossbred lactating cows to evaluate the production economics of supplemental bypass fat and niacin. The treatment group received the same ration with additional 200g bypass fat and 12g niacin daily. The blood glucose, serum cholesterol, triglycerides, BUN, creatinine, total protein, AST, ALT and GGT concentration were within the physiological range among the two groups. There was no influence of the dietary treatment on the plasma levels of creatinine (0.92 vs 0.92 mg/dl), triglycerides (37.40 vs 38.40), total protein (7.18 vs 7.46). At 120 days the glucose level was significantly ($P<0.01$) higher in supplemented group. Low glucose level in high yielding animals during early lactation predisposes animals to ketosis and supplementation of bypass fat and niacin was effective in improving the blood glucose level. The BUN was significantly ($P<0.01$) reduced with the supplementation (14.60±0.46 vs. 22.10±0.71 mg/dl). The total cholesterol (138.44 vs 176.6 mg/dl) was significantly ($P<0.01$) increased with fat and niacin supplementation.

2.36 Economics of Protected Fat and Protein Feeding

Shelke *et al.* (2011b) studied economics of feeding of lactating buffaloes with rumen protected protein and fat which showed that net return over feed cost of milk yield per animal per day was Rs.177.50 and 210.01 in control group and treatment group, respectively, indicating that it was higher by Rs. 32.51 in treatment group over that of control group. Similarly, net return over feed cost of 6% FCM yield per animal per day in control and treatment group was Rs.177.45 and 224.36, respectively, indicating that it was higher by Rs. 46.91 in treatment group over that of control group.

Vahora *et al.* (2013) conducted on-farm feeding trial at the Jalat and Borkheda villages of Dahod district in Gujarat state for 150 days to know the effect of feeding formaldehyde treated *guar* meal (*Cyamopsis tetragonoloba*) and commercial bypass fat supplement to recently calved 18 Mehsani buffaloes in their 2nd to 3rd lactation belonging to different farmers. The buffaloes yielding 8–10 kg milk/day were randomly allotted to three dietary treatments (6 in each group) i.e. T_1 (Control) and T_2 (Bypass protein) and T_3 group (Bypass protein and bypass fat), based on daily milk yield, fat % and body weight and offered isonitrogenous and isocaloric ration. The daily feed cost was found to be lower by 1.68% in T_2 and higher by 6.99% in T_3 over T_1. The daily realizable receipt from sale of milk (Rs./head) was higher ($P<0.01$) in T_3 & T_2 than T_1, which was due to higher fat yield in T_2 and T_3. The daily return over feed cost (ROFC) was higher ($P<0.01$) in T_2 and T_3 as compared to T_1 group.The ROFC was found to be higher by 39.68% under T_2 and 42.75% under T_3 over control ration (T_1). However, it was found to be less by 3.07% in T_2 over T_3 group.

Gajera *et al.* (2013) evaluated the effect of incorporation of rumen protected lysine, methionine and fat on nutrient utilization, nutrient intake and digestibility coefficient of nutrients in Jaffrabadi buffalo heifers. The results revealed that average daily feed cost was Rs. 72.94, 88.34, 92.94 and 109.40 per animal, respectively in T_1 (Control), T_2 (bypass fat), T_3 (bypass lysine & methionine) and T_4 (bypass fat plus bypass lysine & methionine). The cost of feeding was found to be higher in T_4 than that in T_1, T_2 and T_3, the treatment differences were found to be significant ($P<0.05$).

Grewal *et al.* (2014) conducted 120 days field study on crossbred lactating cows to evaluate the production economics of supplemental bypass fat and niacin. The treatment group received the same ration with additional 200g bypass fat and 12g niacin daily. The cost of feed was Rs.147/d in control and it increased by Rs. 24.60/d on supplementation of fat (200 g/d) and niacin (12 g/d) in the treatment group. Higher milk yield in treatment group (1.55 kg/d) resulted in higher value of milk produced in treatment group (Rs. 387.32 vs. 351.637) than control. Thus the price of additional milk produced by treatment group animals resulted in a net daily profit of Rs. 11.05 per animal over the control group due to bypass fat and niacin supplementation.

Chapter 3

Material and Methods

The present experiment entitled, "Protected Nutrient Technology for Crossbred Cattle" was conducted at Research Cum Development Project on Cattle, Department of Animal Husbandry and Dairy Science, Mahatma Phule Krishi Vidyapeeth, Rahuri Dist. Ahmednagar, India located at 19° 23' 0" N and 74° 39' 0" E at an altitude of 511 mtr. Minimum and maximum ambient temperature range from 10 to 20°C in winter and 30 to 40°C in summer with annual rainfall of 561.6 mm.

3.1 Selection and Distribution of Animals

Twenty four crossbred cows in second to fourth lactation with most probable production ability (MPPA) of average approximately 2300 liter milk production per lactation for each group were selected from the herd maintained at RCDP on Cattle. The experimental animals were randomly divided into four treatment groups of 6 in each (Table 1).

Table.1: MPPA of Experimental Animals

Treatment	T_0	T_1	T_2	T_3
1	2899	2847	2701	2853
2	2502	2755	2416	2785
3	2279	2214	2353	2221
4	2235	2104	2254	2066
5	1997	1915	2065	2057
6	1792	1870	1960	1800
Mean	**2284**	**2284.17**	**2291.5**	**2297**
SE	103.14			
CD @5	NS			
CV%	11.04			

The Most Probable Production Ability (MPPA) or Expected Producing Ability (EPA) was computed on the basis of formula given by Lush (1945) as follows:

$$MPPA = A + (P-A) \frac{n\,r}{1 + (n-1)\,r}$$

Where A – Population mean

n – Total number of animals

r – Repeatability of lactation milk record

P – Milk yield in previous lactation

3.2 Treatment Details

The duration of experiment was considered from 30 days prepartum to 90 days postpartum. The animals were fed according to the nutrient requirement of livestock and poultry laid down by ICAR 1998 i.e @ 3.0 % DM of their body weight. Fresh and clean drinking water has made available throughout the experimental period. The measured quantity of roughages i.e. chaffed whole Sugarcane, maize green followed by dry jowar kabdi was offered after the concentrate feeding and in the afternoon, green lucerne and jowar kadbi was offered to experimental animals. 2.0 kg concentrate mixture was given for body maintenance plus for production allowance i.e 30 per cent of milk production.

Table 2: Treatment Details

Sr. No	Treatment	Feeding source
1	T_0	2/3 DM through roughages (2/3 from dry roughages and 1/3 from green roughages) + 1/3 DM from concentrate mixture (Conc. Mix includes 20% GNC (untreated)
2	T_1	2/3 DM through roughages (2/3 from dry roughages and 1/3 from green roughages) + 1/3 DM from concentrate mixture (Conc. Mix includes 20% GNC treated with formaldehyde (FA) @ 1.0 g FA/ 100g CP)
3	T_2	2/3 DM through roughages (2/3 DM from dry roughages and 1/3 DM from green roughages) +1/3 DM from concentrate mixture (Conc. Mix includes 20% GNC untreated) + Bypass Fat(99%) @ 10g/lit milk production)
4	T_3	2/3 DM through roughages (2/3 DM from dry roughages and 1/3 DM from green roughages) +1/3 DM from concentrate mixture (Conc. Mix includes 20% GNC treated with formaldehyde (FA) @ 1.0 g FA/100g CP) + Bypass Fat(99%) @ 10g/lit milk production)

3.3 Preparation of Protected Protein

Groundnut cake was purchased from the market. The cake was first crushed in feed mill and passed through 2.5 mm sieve size. Then the cake was treated with formalin (40% formaldehyde) at the rate of 1.0 g formaldehyde/ 100 g CP of cake. After treatment, the cake was mixed thoroughly and then finally stored in tightly sealed plastic bags for at least 4 to 5 days, as per the procedure standardized for the protection of protein cake to make it bypass protein (Chatterjee and Walli, 2003).

Treatment of formaldehyde (FA) to GNC

Level of Formaldehyde (40%) @ 1.0 g FA/100 g CP

= 2.5 ml formaldehyde/100 g CP

i.e = 25 ml formaldehyde/kg CP

CP content of GNC = 38 % (on dry matter basis)

1 kg CP required 25ml formaldehyde so

100 kg GNC (38 kg CP) need 950 ml formaldehyde (40 %).

3.4 Protected Fat

The Rumen protected fat of palm fatty acids was purchased from Vetcare, Provimi Nutrition Division, Division of Tetragon Chennai Pvt. Ltd, Yelahanka Town, Bangalore, India.

3.5 Observations Recorded

3.5.1 Pre partum study

3.5.1.1 Feed intake

The experimental feeding started before 30 days from expected date of calving of animal. Daily DM intake was observed by recording the daily feed offered and leftover. before 30 days from expected date of calving of animal. The DM of different feed ingredients and concentrate was estimated once every week. Daily intake of nutrients of each cow was recorded throughout the experimental period.

3.5.1.2 Body weight

Body weight of the animals was recorded at fortnightly interval by Avery platform balance. The animals were weighed in the morning before offering feed and water.

3.5.1.3 Body condition score

To assess the body condition of the animal with a fairly high accuracy, a technique developed by Prasad (1994) was adopted in the present study. Body condition score of control and experimental groups of animals were observed on every fortnight during experimental period.

3.5.1.4 Analysis of blood

Blood samples were collected at fortnightly interval up to calving of animals. The blood samples from individual animals were collected by puncturing jugular vein. The serum were separated by centrifugation of the blood samples at 5000 RPM for 5 min and stored in deep freeze for subsequent analysis by using fully automatic blood analyzer Miura 200 (Germany made) for following parameters.

Albumin - (BCG dye method-end point)

Total protein - (Biuret method-end point)

Cholesterol- Chod-pod method end point

Triglycerides- Gpo-pod method end point

Uric acid - Uricase-pod method end point

Creatinine- Modified Jaffe's method

Glucose - God-Pod method end point

NEFA- Colorimetric method

Blood urea nitrigen UV (Gldh/UV – kinetic method)

3.5.1.5 Parturition Related Parameters

3.5.1.5.1 Calf weight at the time of birth

The body weights of the calves born of experimental cows were recorded immediately after birth using Avery Platform balance. The help of veterinary gynecologist was taken for per-rectal examination to check the complete involution of uterus.

3.5.2 Postpartum study

3.5.2.1 Lactation study

3.5.2.1.1 Milking of animals

Animals were milked twice a day i.e. early in morning (5.00 AM) and in the evening (5.00 PM).

3.5.2.1.2 Body weight

Body weight of the animals was recorded at fortnightly intervals by using Avery Platform balance. The animals were weighed in the morning before offering feed and water.

3.5.2.1.3 Body condition score

The body condition score of control and experimental groups of animals were noted on every fortnight up to end of experiment.

3.5.2.1.4 Feed intake

Daily DM intake was observed by recording the daily feed offered and left over throughout the experimental period of 30 days pre-partum to 90 days post-partum. Daily intake of nutrients of each cow was recorded throughout the experimental period.

3.5.2.1.5 Milk yield

Daily milk yield for individual animals were recorded by making the sum of both time (morning and evening) milking.

3.5.2.1.5.1 Calculation for fat corrected milk (FCM)

For the conversion of whole milk into 4 per cent FCM, the following equation was adopted. (Rice *et.al.* 1970)

$$4\ \%\ \text{FCM (kg)} = 0.4 \times M + \frac{15 \times F \times M}{100}$$

Whereas, M = Milk yield (kg)

F = Milk fat (%)

3.5.2.1.5.2 Calculation for energy corrected milk (ECM)

Energy corrected milk yield was calculated by applying the equation by Sjaunja *et al.* (1990).

ECM = (milk production* (0.383*% fat + 0.242* % protein + 0.7832)/ 3.1138

3.5.2.1.6 Milk composition

Milk samples from individual animals were collected and analyzed for milk composition at fortnightly intervals throughout the experimental period. The samples were collected twice (morning and evening) from each animal and were analyzed for its composition by using pre calibrated Milk Analyzer for two days and average of two days was taken for milk composition (Lactostar, FUNKE GERBER, Article No 3510, Berlin).

3.5.2.1.7 Fatty acid analysis of milk

From the pooled milk samples of individual animal, during early lactation the fatty acid analysis of milk was done by using NIR Spectrometer make Zeutec Rendsbory/Germany 2005.

3.5.2.1.8 Analysis of blood

Blood samples were collected at monthly intervals up to 90 days of parturition. The blood samples from individual animal were collected from jugular vein. The serum was separated by centrifugation at 5000 RPM for 5 minutes and stored in deep freeze for subsequent analysis of albumin, glucose, non esterified fatty acids (NEFA), triglycerides, blood urea nitrogen (BUN), Blood cholesterol, creatinine and uric acid as described during pre partum study.

3.5.2.2 Digestibility study

A digestion trial of seven days collection period was conducted at the end of experiment. Representative samples of fodder feed offered, residues and faeces were taken every day from individual animal for dry matter determination and pooled samples of 7 days were analyzed after grinding for proximate principles as per A.O.A.C., 2005.

3.5.2.3 Reproduction study

The following parameters were recorded/ estimated during reproduction studies

A. First estrous after calving

B. Service period

C. Number of AI/ conception

D. Conception rate

Cows under study were critically observed 3 times every day for signs of estrus and were inseminated thereafter continuously up to confirmation of pregnancy diagnosis.

3.6 Statistical Analyses

The data generated during experimental period was subjected to statistical analysis, which was done by Factorial Randomized Block Design (FRBD)/ SAS., (2011) 9.3.1. version by following standard method of analysis of variance given by Snedecor and Cochran (1989).

Chapter 4

Results and Discusion

4.1 Composition of Concentrate Mixture

The ingredient composition of experimental concentrate mixture is presented in Table 3. Groundnut cake in T_1 and T_3 were treated with formaldehyde as a tool of protein protection while protected fat at the rate of 10g per liter milk production per animal in T_2 and T_3 was added in concentrate mixture before feeding every day.

Table 3: Ingredient Composition (%) of Concentrate Mixture

Sr. No.	Feed Ingredients	T_0	T_1	T_2	T_3
1.	Rice bran	35	35	35	35
2.	Wheat bran	15	15	15	15
3.	Groundnut cake (Untreated)	20	-	20	-
4.	Groundnut cake (Treated with FA)	-	20	-	20
5.	Tur chuni	27	27	27	27
6.	Mineral mixture	02	02	02	02
7.	Common salt	01	01	01	01
	Total	**100**	**100**	**100**	**100**
1.	Protected Fat (g per litre milk production)	-	-	10	10

4.2 Proximate Composition of Feed Ingredients and Concentrate Mixture

The chemical composition of concentrate mixture used in experiment is presented in table 4. The concentrate mixture of control group (T_0) contained 88.15, 18.61, 13.41, 2.57, 7.85 and 57.56 per cent DM, CP, CF, EE, TA and NFE on the DM basis, respectively. The corresponding values for T_1, T_2 and T_3 were 88.10, 18.65, 13.45, 2.55, 7.85 and 57.50 and 88.00, 18.40, 13.65, 2.62, 7.86 and 57.47 and 88.20, 18.55, 13.71, 2.61, 7.86 and 57.27 per cent, respectively. The proximate composition of different fodders and feed ingredients is given in Table 5.

Table 4: Proximate composition of experimental concentrate mixture (% DM basis)

Parameter	(T_0)	(T_1)	(T_2)	(T_3)
DM	88.15	88.10	88.00	88.20
CP	18.61	18.65	18.40	18.55
CF	13.41	13.45	13.65	13.71
EE	2.57	2.55	2.62	2.61
TA	7.85	7.85	7.86	7.86
NFE	57.56	57.50	57.47	57.27

Table 5: Chemical composition of feed ingredients (% DM basis)

Parameter	Maize	Lucerne	S. Cane	Dry Jowar	Rice bran	Wheat bran	Tur chuni	GNC
DM	17.76	23.73	33.93	60.68	90.72	93.36	92.08	92.06
CP	09.85	21.54	05.24	04.94	15.44	15.07	10.46	38.01
CF	31.90	25.29	31.66	33.74	12.37	04.26	19.99	15.24
EE	01.68	01.85	0.68	0.96	01.13	01.85	01.69	07.21
TA	08.94	11.26	08.18	11.84	12.61	01.89	09.99	02.31
NFE	47.63	40.06	54.24	48.52	58.45	76.93	57.87	37.23

The values of per cent DM, CP, CF, EE, TA and NFE for maize green fodder was 17.76, 09.85, 31.90, 1.68, 8.94 and 47.63 per cent, Lucerne was 23.73, 21.54, 25.29, 1.85, 11.26 and 40.06 per cent, sugar cane was 33.93, 5.24, 31.66, 0.68, 8.18 and 54.24 and dry jawar was 60.68, 4.94, 33.74, 0.96, 11.84 and 48.52 per cent, respectively. The chemical composition of fodders was within normal range (Ranjhan, 1998).

Table 6 : Fatty acid profile (%) of experimental protected fat

Sr. No	Parameters	Per cent
1	Myristic acid	1.27
2	Palmitic acid	79.4
3	Stearic acid	4.97
4	Oleic acid	13.28
5	Linoleic acid	1.03
6	Saturated fat	85.66
7	Mono unsaturated fat	13.28
8	Poly unsaturated fat	1.03

The fatty acid profile of experimental protected fat is presented in table 6, which contain 85.66 per cent saturated fat, 13.28 per cent mono unsaturated fat and 1.03 per cent poly unsaturated fat.

4.3 Body Weight of Experimental Animals during Experimental Period

Average initial body weight (Table 7) of experimental animals were 438.33, 441.67, 405.00 and 458.33 kg in T_0, T_1, T_2 and T_3, respectively with non significantl differences among the treatments.

Table 7: Initial body weight of experimental animals in different treatments

Parameter	T_0	T_1	T_2	T_3
1	480	510	435	550
2	365	440	340	485
3	470	360	445	485
4	450	470	440	440
5	465	520	450	430
6	400	350	320	360
Mean	**438.33**	**441.67**	**405.00**	**458.33**
SE	18.01			
CD @ 5%	NS			
CV %	10.12			

4.4 Body Weight Changes During Experimental Period

Fortnightly changes in body weights of experimental animal during the experimental period are presented in Table 8 and 9. In first fortnight (-30 to -15 days) the experimental animals gained 18.33, 20.83, 20.00 and 19.17 kg body weight in groups T_0, T_1, T_2 and T_3 respectively, while at time of calving experimental animals lost -31.67, -32.50, -28.33 and -29.50 kg body weight in groups T_0, T_1, T_2, and T_3, respectively compared to -15 day body weight. There was net loss of 13.33 kg in T_0 and weight gain of 22.00 kg followed 15.00 and 04.17 kg in T_3, T_2 and T_1 group respectively after 6th fortnight period of calving compare to body weight at calving (Table 8). Body weight change (Table 9) in groups T_3 and T_2 revealed gain in BW from second fortnight onwards as was evident from increasing trend in body weight while it took longer (3rd fortnight) time in case of group T_1 and T_0 (4th fortnight) Feeding protected fat and protein showed effective trend in term of reducing the extent and duration of body weight loss as compared to control group.

These findings concurs with the reports of Grummer *et al.* (1995) who reported higher post-partum body weight of heifers fed on high energy diet pre-partum and supplemental fat (2.8%) post-partum than those fed no supplemental fat. McNamara *et al.* (2003) observed that on supplementation of two protected fat supplements Megalac Plus (0.4 kg/d) containing Ca salts of methionine hydroxyl analogue and Megapro gold (1.5 kg/d) containing Ca salts of palm fatty acids, extracted rapeseed meal and whey permeat to Holstein Friesian cows, there was less change in body weight during early lactation in supplemented groups. Tyagi *et al.*, (2009a) reported that there was less change in per cent body weight on supplementation of Ca salts of fatty acids to cows. Thus, the feeding of supplemental fat to lactating cows was beneficial in improving their post-partum body weights. Garg and Mehta (1998) observed that due to bypass fat feeding, reduction in weight loss in the first quarter was observed.

Table 8: Pre and Postpartum fortnightly changes in body weights (kg) of cattle fed with or without protected fat and protein

Days prepartum	T_0	T_1	T_2	T_3	Mean
-30	438.33	441.67	405.00	458.33	**435.83^a**
-15	456.67	462.50	425.00	477.50	**455.42^a**
0 at Calving	425.00	430.00	395.00	448.00	**424.50^b**
Mean	**440.00^a**	**444.72^a**	**408.33^b**	**461.28^a**	**438.58**
	S	T	SxT		
SEm	11.577	13.368	23.154		
CD5%	23.263	26.862	N/A		
Days postpartum					
0 at Calving	**425.00**	**430.00**	**395.00**	**448.00**	**424.50**
15	410.00	417.50	383.33	442.50	**413.33**
30	403.33	418.33	390.83	450.00	**415.63**
45	414.17	415.00	392.50	450.00	**417.92**
60	408.33	420.00	395.83	453.33	**419.38**
75	409.17	425.83	400.00	461.67	**424.17**
90	411.67	434.17	410.00	470.00	**431.46**
Mean	**411.67^{bc}**	**422.98^b**	**395.36^c**	**453.64^a**	**420.91**
Weight gain after calving to 90 days					
		-13.33^c	**4.17^b**	**15.00^{ab}**	**22.00^a**
	S	T	SxT		
SEm	11.084	8.379	22.169		
CD5%	N/A	16.59	N/A		

abc bearing different superscripts within row and coloum different significantly (P<0.5)

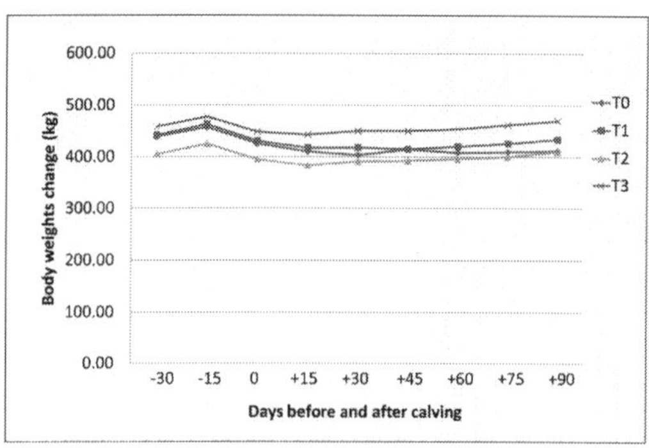

Fig. 1: Pre and Postpartum fortnightly changes in body weights (kg) of cattle fed with or without protected fat and protein

Mathew and Zachariah (2013) reported that supplementation of 100 grams of bypass fat was found to be effective in preventing weight loss during initial 90 days of lactation. Naik *et al.* (2009b) reported better recovery in BW (-2.08 vs +14.13, kg) in crossbred cows during early lactation in bypass fat supplemented group. Thakur and Shelke (2009) reported that supplementation of calcium salts of soya oil fatty acids at 4% of DMI improved the ADG (553.10 vs 577.60, g) in Murrah buffalo calves owing to higher TDN intake (2.14 vs 2.42, kg/d).

Table 9: Pre and Postpartum fortnightly changes in body weight gain/losses (kg) of cattle fed with or without protected fat and protein

Days prepartum	T_0	T_1	T_2	T_3	Mean
-15	18.33	20.83	20.00	19.17	**19.58**[a]
0 at Calving	-31.67	-32.50	-28.33	-29.50	**-30.50**[b]
Mean	**-6.67**	**-5.83**	**-4.17**	**-5.17**	**-5.46**
	S	T	SxT		
Sem	2.452	3.468	4.904		
CD5%	4.998	N/A	N/A		
Days postpartum					
0 at Calving	**-31.67**	**-32.50**	**-28.33**	**-29.50**	**-30.50**[a]
15	-15.00	-12.50	-11.67	-5.50	**-11.17**[b]
30	-6.67	**0.83**	**7.50**	**7.50**	**2.29**[c]
45	**10.83**	-3.33	1.67	0.00	**2.29**[c]
60	-5.83	5.00	3.33	3.33	**1.46**[c]
75	**0.83**	5.83	4.17	8.33	**4.79**[c]
90	2.50	8.33	10.00	8.33	**7.29**[c]
Mean	**-6.43**	**-4.05**	**-1.90**	**-1.07**	**-3.36**
	S	T	SxT		
Sem	3.699	2.796	7.398		
CD5%	7.323	N/A	N/A		

abc bearing different superscripts within row and coloum different significantly (P<0.5)

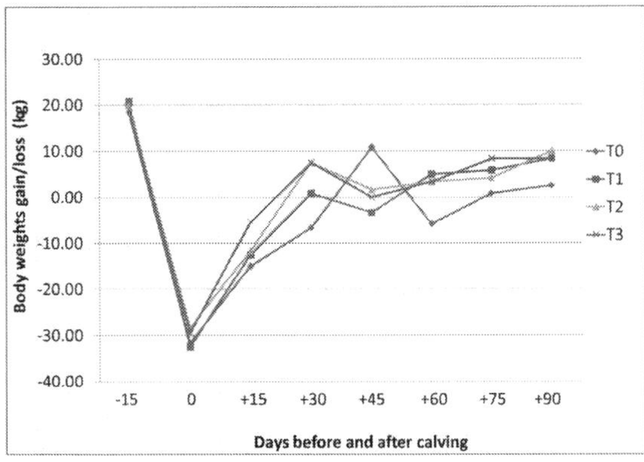

Fig. 2: Pre and Postpartum fortnightly changes in body weights gain/ loss (kg) of cattle fed with or without protected fat and protein

Wadhwa *et al.* (2012) reported the BW of the animals improved considerably in the bypass fat supplemented group as compared to the un-supplemented group (551 vs 508, kg), but the differences were non-significant. Gowda *et al.* (2012) reported less body weight loss in cows fed protected fat and the regain of body weight was much quicker, also Yadav *et al.* (2015) repoted less body loss due feeding prilled fat to crossbed cow. Whereas, Lounglawan *et al.* (2007) and Singh *et al.* (2014) reported similar body loss by feeding bypass fat.

Kalbande and Thomas (1999) reported more fortnightly body weight gain by feeding bypass protein in crossbred cows. Sai *et al.* (2014) reported significantly higher average daily gain by supplementation of rumen protected methionine and lysine in calves.

Shelke *et al.* (2012) reported significantly higher body weight gain by feeding bypass fat and protein in lactating buffaloes. Also Grewal *et al.* (2014) observed more body weight gain by supplementation of bypass fat and niacin in cows.

4.5 Body Condition Score (BCS) Changes

Fortnightly changes in BCS are presented in Table 10. Higher BCS observed in T_3 (3.64) followed by T_2 (3.60), T_1 (3.57) and T_0 (3.48), respectively which was significant ($P<0.05$) higher in T_3, T_2 and T_1 over T_0. From the results it was revealed that feeding of protected fat and protein alone or in combination had significantly beneficial effect on body condition score of the animals. The body condition score of the animals significantly differed with each other during different periods. The body condition score increased in 30 days prepartum period than 15 days prepartum period. The body condition score of the animals decreased significantly in 15 days after calving and it again showed continuous increasing trend up to the 90 days period after calving. After calving, animals reached at highest body condition score which indicated that whatever the calving stress observed in animals was nullified in 90 days period. Therefore, from these results it is revealed that the animals were going in stress after calving, feeding bypass fat and protein has positive effect and is essential to recoup the animals within 90 days period after calving.

Table 10: Fortnightly changes in Body Condition Score (BCS) of cattle fed ration with or without protected fat and protein

Treat	-30	-15	0	15	30	45	60	75	90	Mean
T_0	3.79	4.04	3.29	3.08	3.13	3.21	3.38	3.63	3.79	**3.48[c]**
T_1	3.75	4.04	3.54	3.17	3.25	3.29	3.54	3.71	3.83	**3.57[b]**
T_2	3.79	4.04	3.50	3.21	3.29	3.29	3.58	3.79	3.92	**3.60[ab]**
T_3	3.83	4.08	3.50	3.21	3.33	3.42	3.63	3.83	3.92	**3.64[a]**
Mean	**3.79[c]**	**4.05[a]**	**3.46[e]**	**3.17[g]**	**3.25[f]**	**3.30[f]**	**3.53[d]**	**3.74[c]**	**3.86[b]**	**3.57**
	S	T	SxT	CV%						
SEm	0.022	0.015	0.045							
CD5%	0.062	0.041	NS	3.407						

abc bearing different superscripts within row and coloum different significantly (P<0.5)

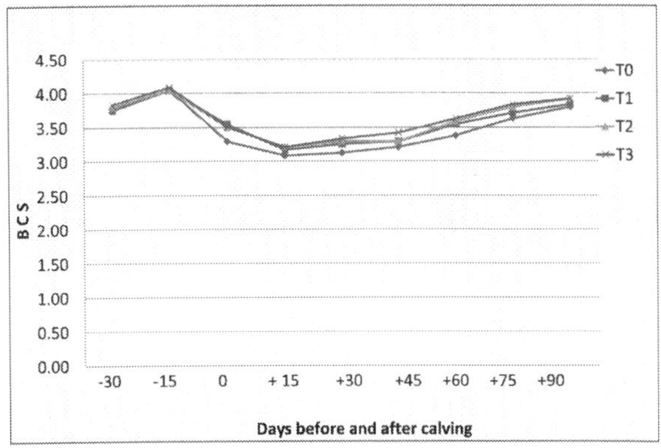

Fig. 3: Fortnightly changes in Body Condition Score (BCS) of cattle fed ration with or without protected fat and protein

Body condition score (BCS) provides better estimate of body fat distribution than body weight (Ferguson *et al.*, 1994). Treacher *et al.* (1986) reported general pattern of change in BCS over lactation is an initial fall continuing for 2-3 months and then a lower recovery over mid lactation. The present findings are in agreement with the findings of Grummer *et al.* (1995) who reported higher post-partum body condition scores of heifers fed on high energy diet pre-partum and supplemental fat (2.8%) post-partum than those fed no supplemental fat. Garg and Mehta (1998) reported improvement in body condition score of cows fed supplemental bypass fat fed group and deterioration in that of control group. Methew and Zachariah (2013) observed that supplemented bypass fat showed better BCS than control. Naik *et al.* (2009b) reported better recovery in BSC (-0.06 vs +0.02) in crossbred cows during early lactation in bypass fat supplemented group. Whereas, Lounglawan *et al.* (2007) and Singh *et al.* (2014) reported decline in BCS (P<0.01) was more in control group cows than the prill fat cows group. Kirovski *et al.* (2015) concluded that the loss in body condition was significantly lower in the group fed palm oil than in the control group cows. Yadav *et al.* (2015) reported significantly inscresed BCS by feeding prilled fat to crossbed cows.

4.6 Calving Performance

Average birth weight of calves born of cows fed rations with or without rumen protected fat and protein are presented in Table 11. From the table it is revealed that, avarege calf birth weight was higher in T_3 (25.5 kg) followed by T_1 (23.83 Kg), T_0 (23.17) and T_2 (20.67kg), which was statistically non-significant among treatment. The results indicated that prepartum feeding of either protected protein or fat or combination of both was unable to exert its effect on birth weight of calves born. These findings are not in agreement with the findings of Tyagi *et al.* (2009[a]) and Shelke *et al.* (2012a). They reported significantly increase in birth weight of calve by feeding bypass fat and protein to cows.

Table 11: Calf birth weight in cattle fed ration with or without rumen protected fat and protein (kg)

Treatment	T_0	T_1	T_2	T_3
1	18	30	18	26
2	24	18	22	41
3	20	24	22	22
4	28	21	24	19
5	29	28	17	23
6	20	22	21	22
Average	**23.17**	**23.83**	**20.67**	**25.5**
Mean	23.29			
SE	2.34			
CD @ 5%	NS			
CV%	24.58			

4.7 Effect of Protected Fat and Protein Supplementation on Nutrient Digestibility

The overall digestibility of DM, CP, CF, EE and NFE observed in present investigation was 58.87 ± 1.396, 63.72 ± 1.275, 65.61 ± 1.134, 66.33 ± 0.960 and 60.34 ± 1.303 per cent in T_0, T_1, T_2 and T_3, respectively (Table 12).

The digestibility coefficient of DM was 58.49, 56.51, 59.43 and 61.04 per cent in T_0, T_1, T_2 and T_3 treatment, respectively. The protected fat and protein supplementation had no significant effect on digestibility of dry matter. These results are in agreement with the findings of Naik *et al.* (2007b; 2009a); Purushathaman *et al.* (2008); Tyagi *et al.* (2009a); Thakur and Shelke, (2012a); Sirohi *et al.* (2010); Naik (2013a); Sonttakke *et al.* (2014) and Yadav *et al.*

(2015) who reported no effect of supplementation of bypass fat on the digestibility of DM which may be due to the non-interference and relatively stable nature of bypass fat (Garcia-Bojalil *et al.*, 1998). However, Sarwar *et al.* (2003) reported decreased digestibility of DM when bypass fat was fed to the dairy animals. Also Kumar *et al.* (2005); Yadav and Choudhary (2010); Arewad *et al.* (2011); Sirohi *et al.* (2013) and Sai *et al.* (2014) reported no effect of supplementation of bypass protein on the digestibility of DM. However, Wankhede and Kalbande (2001); Mishra *et al.* (2006); Torane *et al.* (2006): Patel *et al.* (2012): Gajera *et al.* (2013) and Movaliya *et al.* (2013) reported increase in the digestibility of DM when bypass protein was fed to the animals. Whereas Shelke *et al.* (2011); Grewal *et al.* (2014) and Nam *et al.* (2014) reported no effect of bypass fat and protein on the digestibility of DM however, Gajera *et al.* (2013) reported increase in the digestibility of DM when fed bypass fat and protein.

The digestibility coefficient of crude protein was 62.04, 64.50, 63.12 and 65.20 per cent in T_0, T_1, T_2, and T_3, respectively. The treatment had non significant influence on the digestibility of crude protein. The supplemented rumen protected fat and protein did not affect the digestibility of CP which might be due to the amount of free fat in the rumen was very small to affect the microbial population to a large extent. The bypass fat with higher extent of protection or with lower amount of supplementation does not affect the CP digestibility in ruminants. The results in present study are in agreement with those of Naik *et al.*, (2007b); (2009a); Purushathaman *et al.* (2008); Tyagi *et al.*, (2009a); Thakur and Shelke (2010); Sirohi *et al.* (2010); Naik (2013a); Sonttakke *et al.* (2014) and Yadav *et al.* (2015) who reported no effect on CP digestibility by feeding different bypass fat to lactating cows and buffaloes. However, Sarwar *et al.* (2003) reported decreased in the digestibility of CP when bypass fat was fed to the dairy animals while Schauff and Clark (1992) reported increase in the digestibility of CP, when Ca-LCFA was fed to the dairy animals.

From the results obtained in the present study, it may be interpreted that formaldehyde treatment at the level of 1.0 g/100g CP did not overprotect the groundnut cake, and thus the CP digestibility remained similar in both the groups. Similarly Tiwari and Yadav (1994); Chatterjee and Walli (1998); Walli and Sirohi (2004); Kumar *et al.* (2005); Mishra *et al.* (2005); Yadav and Choudhary (2010); Arewad *et al.*(2011)and Sirohi *et al.* (2013) found that CP digestibility was not significantly affected by feeding of bypass proteins. However, Wankhede and Kalbande (2001); Torane *et al.* (2006); Yadav *et al.* (2006); Patel *et al.* (2012); Movaliya *et al.* (2013) and Sai *et al.* (2014) reported increase in the digestibility of CP, when different bypass proteins were fed to the dairy animals. Also similar results were obtained by Shelke *et al.* (2012a) and Grewal *et al.* (2014) by feeding bypass fat and protein whereas Gajera *et al.* (2013) reported increase in the digestibility of CP, when fed bypass fat and protein.

The crude fiber digestibility was significantly (P<0.05) influenced by feeding of protected fat and protein supplementation. The treatment T_1 (66.55), T_2 (66.18) and T_3 (67.27) were at par but had significantly higher crude fiber digestibility than control T_0 (62.43) treatment. It might be due to protected fat and protein which reduce energy losses associated with fermentation and get more availability of energy to cellulolytic bacteria which increase fibre digestion. These findings concur with the findings of Wankhede and Kalbande (2001); Torane *et al.* (2006); Yadav *et al.* (2006); Patel *et al.* (2012) and Movaliya *et al.* (2013). However, Naik *et al.* (2007b; 2009a); Tyagi *et al.* (2009a); Thakur and Shelke (2010); Sirohi *et al.* (2010) and Naik (2013a) reported no effect of supplementation of bypass fat on the digestibility of CF. Similarly, Kumar *et al.* (2005); Mishra *et al.* (2005); Yadav and Chaudhary (2010); Arewad *et al.*(2011) and Sirohi *et al.* (2013) found that CF digestibility was not significantly affected by feeding of bypass proteins. Gajera *et al.* (2013) reported increased CF digestibility by feeding bypass protein and fat whereas Shelke *et al.* (2012a) reported no significant difference by feeding bypass protein and fat.

Table 12: Digestibility coefficients of different nutrient by feeding protected fat and protein

Nutrients	T0	T1	T2	T3	Mean	SE	CD@5	CV%
DM	58.49	56.51	59.43	61.04	58.87	1.396	NS	5.810
CP	62.04	64.50	63.12	65.20	63.72	1.275	NS	4.900
CF	62.43b	66.55a	66.18a	67.27a	65.61	1.134	3.418	4.234
EE	62.07b	66.59a	68.41a	68.24a	66.33a	0.960	2.893	3.545
NFE	58.52bc	57.17c	62.19ab	63.46a	60.34	1.303	3.927	5.289

abc bearing different superscripts within row and coloum different significantly (P<0.5)

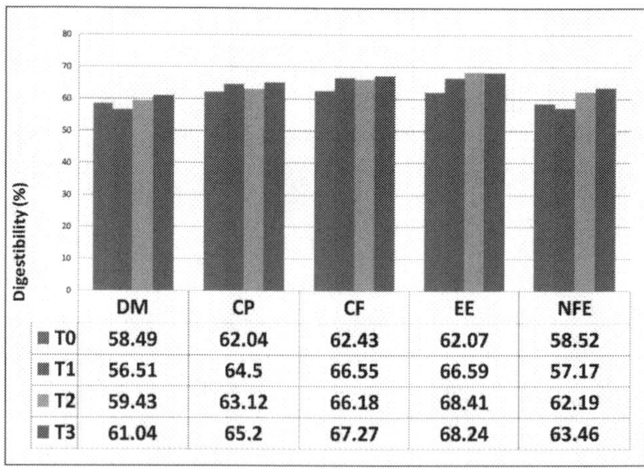

Fig. 4: Digestibility of coefficients of different nutrient by feeding protected fat and protein

The digestibility coefficient of EE was significantly ($P<0.05$) higher in T_2 (68.41), T_3 (68.24), T_1 (66.59) than T_0 (62.07), respectively. The treatments T_2, T_3, T_1 were at par and significantly higher over control T_0. In present study EE digestibility coefficient indicated that feeding of protected protein and fat alone or in combination increases the EE digestibility which might be due to high quality fat inclusion in experimental group. The results in present study are in agreement with those of Palmquist (1991); Naik *et al.*, (2007b; 2009a), Purushathaman *et al.* (2008), Thakur and Shelke (2010), Sirohi *et al.* (2010); Sonttakke *et al.* (2014) and Yadav *et al.* (2015) who reported significantly increased EE digestibility by feeding different bypass fat to lactating cows and buffaloes. Similar results were obtained by Wankhede and Kalbande (2001); Yadav *et al.* (2006); Mishra *et al.* (2006); Torane *et al.* (2006); Yadav and Chaudhary (2010); Patel *et al.* (2012) and Movaliya *et al.* (2013) who reported significantly increased EE digestibility by feeding different bypass fat. However, Sarwar *et al.* (2003) and Naik (2013a) reported no effect in the digestibility of EE when bypass fat was fed to the dairy animals. Kumar *et al.* (2005); Mishra *et al.* (2005); Arewad *et al.*(2011) and Sirohi *et al.* (2013) found that EE digestibility was not affected by feeding of bypass proteins. However, Shelke *et al.* (2012a); Gajara *et al.* (2013) and Grewal *et al.* (2014) reported significantly increased EE digestibility by feeding different bypass fat and protein feed.

The digestibility coefficients of NFE were 58.52, 57.17, 62.19 and 63.46 per cent in T_0, T_1, T_2 and T_3 respectively. NFE digestibility was highest in cows fed protected fat and protein combination (T_3) followed by protected fat alone (T_2). Treatment (T_3) and (T_2) were at par with each other and treatment T1 had significantly lower NFE digestibility. These findings are in contras with reports of Naik *et al.*, (2007b; 2009a); Tyagi *et al.* (2009a); Thakur and Shelke (2010); Sirohi *et al.* (2010) who reported non significant effect of supplementation of bypass fat on the digestibility of NFE. The findings of Movaliya *et al.* (2013) are in agreement who observed significantly increased NFE digestibility by feeding bypass protein. Whereas, Kumar *et al.* (2005); Mishra *et al.* (2005); Yadav and Chaudhary (2010); Arewad *et al.* (2011); Sirohi *et al.* (2013) and Sai *et al.* (2014) found that NFE digestibility was not affected by feeding of bypass proteins. Gajera *et al.* (2013) reported that NFE digestibility significantly increased by feeding bypass fat and niacin, where as, Shelke *et al.* (2012a) reported no effect on NFE digestibility by feeding bypass fat and protein.

4.8 Effect on Nutrient Intake

Average DM and nutrient intake in different treatment group is presented in Table 13. The DMI was 12.56, 12.59, 12.23 and 12.72 kg/day in T_0, T_1, T_2 and

T_3 respectively, which was significantly ($P<0.05$) higher in T_3 followed by T_1, T_0 and T_2. It might be due to higher body weight in T_3, T_1 and T_0 than T_2. However, from the table it is revealed that the DMI/100 kg body weight was 3.05, 3.01, 3.13 and 2.81 kg/d in T_0, T_1, T_2 and T_3 respectively which did not differ among all groups. The results confirmed the finding that in case of crossbreds, the dry matter requirements is @ 3% of the body weight of animal. Garg and Mehta (1998); Sarwer *et al.* (2003); Purushothaman *et al.* (2008); Naik *et al.* (2007b; 2009a); Tyagi *et al.* (2009b); Thakur and Shelke (2010); Sirohi *et al.* (2010); Zhang *et al.* (2011); Mudgal *et al.* (2012); Singh *et al.* (2014); Sonttakke *et al.* (2014) and Ramteke *et al.* (2014) reported that the DM intake of dairy animals was not altered on supplementation of bypass fat. However, Chouinard *et al.* (1997) reported decrease in DMI while Tyagi *et al.* (2009a) and Fiorentin *et al.* (2012) reported increase in DM intake in dairy animals fed bypass fat. Chatterjee and Walli (2003); Mishra *et al.* (2006); Kumar *et al.* (2005); Arewad *et al.* (2011); Sirohi *et al.* (2013) and Sai *et al.* (2014) observed non significant effect on supplementing bypass protein on DMI. The results of the present study indicated that there was no adverse effect of rumen protected fat and protein supplementation on DMI of lactating crossbreed cattle. Similarly, Garg *et al.*, (2002) reported that by supplementing bypass nutrients (fats encapsulated in a aldehyde treated protein) @ 1000 g/d to lactating cows and buffaloes, there was no effect on DMI. Whereas, Kalbande and Thomas (1999); Wankhede and Kalbande (2001); Yadav *et al.* (2006); Movaliya *et al.* (2013) and Amrutkar *et al.* (2014) reported significantly increase in DMI by feeding bypass protein and also Shelke *et al.* (2012b) and Vahora *et al.* (2013) reported significant increase in DMI by feeding bypass fat and protein.

Table 13: Dry matter and nutrient intake in different treatment groups

Treatment	DMI (kg/day)	DMI/ (%BW)	DCP Intake (kg/day)	DCPI (%BW)	TDN Intake (kg/day)	TDNI /(%BW)
T_0	12.47^a	3.05	1.15^b	0.28	7.11^c	1.74
T_1	12.59^a	3.01	1.21^a	0.29	7.86^b	1.88
T_2	12.23^b	3.13	1.18^a	0.30	7.93^b	2.03
T_3	12.72^a	2.81	1.23^a	0.27	8.32^a	1.84
Mean	**12.50**	**3.00**	**1.19**	**0.28**	**7.80**	**1.87**
SE(+)	0.11	0.12	0.01	0.01	0.07	0.07
CD @5%	0.34	NS	0.03	NS	0.22	NS
CV (%)	2.24	9.42	2.23	9.67	2.24	9.79

abc bearing different superscripts within row and coloum different significantly ($P<0.5$)

The average digestible crude protein intake was 1.15, 1.21, 1.18 and 1.23 kg/d in T_0, T_1, T_2 and T_3 respectively which was significantly higher ($P<0.01$) in T_3, T_2 and T_1 than T_0 due to higher DMI. However, DCPI/ 100 kg body weight were 0.28, 0.29, 0.30 and 0.27 kg/d in T_0, T_1, T_2 and T_3, respectively. There was no difference among all groups.

Average total digestible nutrients intake (TDN) were 7.11, 7.86, 7.93 and 8.32 kg/day in T_0, T_1, T_2 and T_3, respectively which was significantly higher ($P<0.01$) in T_3 followed by T_2 and T_1 over T_0 might be due to supplementation of protected fat and protein. The TDNI/ 100 kg body weight were 1.74, 1.88, 2.03 and 1.84 kg/day in T_0, T_1, T_2 and T_3 treatment, respectively showing the similar trend as that of DMI and DCPI/ 100kg body weight. There were reports of no effect on CP intake [Sarwar *et al.* (2003); Tyagi *et al.* (2009a); (2009b); Thakur and Shelke (2010)] and DCP intake (Naik *et al.*, 2007b) by the supplementation of bypass fat to dairy animals. However, Sirohi *et al.* (2010); Singh *et al.*(2014) and Ramteke *et al.* (2014) reported increase in CP intake on supplementation with bypass fat. The TDN intake was either not altered (Naik *et al.*, 2007b; Sirohi *et al.*, 2010) or increased (Tyagi *et al.*, 2009a; Thakur and Shelke, 2010 and Ramteke *et al.* 2014) on supplementation of bypass fat in the diet of the dairy animals. Also Kumar *et al.* (2005) and Sai *et al.* (2014) reported no significant difference on CP and TDN intake, whereas Amrutkar *et al.* (2014) reported increased in CP and TDN intake by supplementing rumen protected Methionine and lysine. Shelke *et al.* (2012b) and Gajera *et al.* (2013) reported increased in CP and TDN intake by supplementing bypass fat and protein.

4.9 Effect of Protected Fat and Protein Supplementation on Milk Production (Kg/Day) and Milk Composition

4.9.1 Milk Production

Average daily milk yield kg/day (Table 14) ranged from 7.85 to 10.79 kg/d in group T_0, 9.72 to 12.64 kg/day in T_1, 8.71 to 12.48 kg/d in T_2 and 10.03 to 13.55 kg/d in T_3 in different fortnights. The average milk production during supplementation period was 9.82, 11.76, 11.41 and 12.43 kg/d in group T_0, T_1, T_2 and T_3, respectively which was 16.49% higher in T_1, 13.93% higher in T_2 and 20.99% in T_3 as compared T_0. The treatment and period had significant effect on average daily milk yield of the animals. The milk yield from period 15 days had shown increasing trend up to 45 days period but again it showen decreasing trend up to 90 days period. However, milk yield from period 30 days to 90 days were at par with each other indicating the consistency in milk yield over the periods. The results also revealed that milk yield in treatment fed protected protein and fat in combination (T_3) had highest milk yield followed by treatment protected protein alone (T_1) and fat alone (T_2) over the control group (T_3).

Table 14: Fortnightly average milk yield (kg/d) in cattle fed with or without protected fat and protein

Treat	Days							
	15	30	45	60	75	90	Mean	% ↑↓
T_0	07.85	10.79	10.69	10.23	09.81	09.53	**09.82[b]**	
T_1	09.72	11.47	12.64	12.60	11.80	12.33	**11.76[a]**	**16.49**
T_2	08.71	10.84	12.26	11.80	12.48	12.34	**11.41[a]**	**13.93**
T_3	10.03	12.34	13.15	13.55	13.39	12.12	**12.43[a]**	**20.99**
Mean	**09.08[b]**	**11.36[a]**	**12.19[a]**	**12.05[a]**	**11.87[a]**	**11.58[a]**	**11.35**	
	S	T	SxT	CV%				
SEm	0.494	0.403	0.988					
CD5%	1.370	1.118	NS	25.410				

abc bearing different superscripts within row and coloum different significantly (P<0.5)

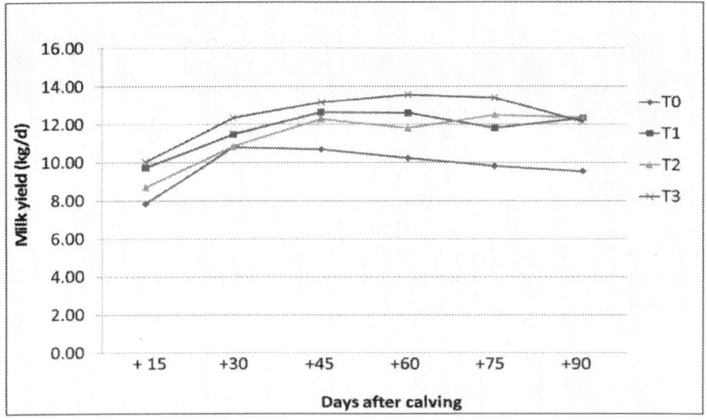

Fig. 5: K Fortnightly average milk yield (kg/d) in cattle fed with or without protected fat and protein

In present study higher degradability of untreated groundnut cake might cause an excessive production of NH_3 in rumen, that may be higher than the quantities needed by rumen microbes for use of microbial protein synthesis. Thus, a significant amount of NH_3 might have got absorbed through rumen wall and then partly lost through urine in the form of urea. The synthesis of urea itself is an energy consuming process and rumen microbial protein production is an energy dependent mechanism. Thus, loss of excess nitrogen in the form of urea not only cause a decreased efficiency of protein utilization but could also cause decreased efficiency of energy utilization. However, when formaldehyde treated protected protein replaced untreated one, this excessive loss of both nitrogen and energy could perhaps was avoided due to the much lower degradability of the protein. This might have resulted in an increased energy and nitrogen balance causing further increase in milk production and the yield of different milk constituents. Also due to protected fat more availability of energy through the fatty acids to mammary gland might have increased the milk yields.

The improved supply of amino acids in the presence of sufficient energy might have also improved the protein-energy balance and created a better balance of precursors for milk synthesis, resulting in increased milk production. Therefore, the milk yield in treatment feeding protected protein and fat in combination (T_3) had highest milk yield followed by treatment protected protein (T_1) and fat alone (T_2) over the control group (T_0). Also supplementation of protected protein and fat before parturition might be reduced the detrimental effects of negative energy balance which might have increased both lactation as well as metabolic performance.

The significant improvement in milk production on supplementation of rumen protected fat and protein was in line with the findings of many researchers. Schneider *et al.* (1988); West and Hill (1990); Sklan *et al.* (1991); Schauff and Clark (1992); Scot *et al.* (1994); Skan *et al.* (1994); Chouinard *et al.* (1997); Gerg and Mehta (1998); Sarwar *et al.* (2003); Lounglawan et al. (2007); Purushothaman *et al.* (2008); Barley and Baghel (2009); Sirohi *et al.*, 2010); Thakur and Shelke, 2010; Tyagi *et al.* (2010); Zhang *et al.* (2011); Gowda *et al.* (2013); Wadhawa *et al.* (2012); Qureshi and Tawheed (2012); Ranjan *et al.* (2012); Mudgul *et al.* (2012); Naik *et al.* (2013); Sontakke *et al.* (2014); Ramteke *et al.* (2014); Kirovski *et al.* (2015) and Yadav *et al.* (2015) were reported significantly increased milk yield by feeding bypass fat.

Positive response on milk production performance as a result of feeding protected proteins and amino acids have been observed by Morgan (1985); Kunju *et al.* (1992); Sampath *et al.* (1997); Garg (1998); Kalbande and Thomas (1999); Gulati *et al.* (2002); Garg *et al.* (2003b); Garg *et al.* (2004); Garg *et al.* (2005); Yadav and Chaudhary (2004); Walli and Sirohi (2004); Sahoo and Walli (2005); Sampath *et al.* (2005); Kumar *et al.* (2005); Trinacly *et al.* (2006); Garg *et al.* (2007b); Mishra *et al.* (2008); Chandrashekharan *et al.* (2008); Garg and Sherasia (2010); Suresh *et al.* (2011); Dosky *et al.* (2012); Sherasia *et al.* (2012); Vahora *et al.* (2012); Sirohi *et al.* (2013); Amrutkar *et al.* (2014) and Kumar *et al.* (2015) who reported significantly increased milk yield by feeding bypass protein. Whereas, Quisi and Titi (2014) reported no effect of rumen protected methionine in goat.

Similarly, Garg *et al.* (2002); Strusinska *et al.* (2006); Shelke *et al.* (2011); Garg *et al.* (2012); Dhulipalla *et al.* (2013); Vahora *et al.* (2013); Grewal *et al.* (2014) and Nam *et al.* (2014) had also reported significantly increased milk yield by feeding bypass fat and protein.

4.9.2 Effect on milk composition

4.9.2.1 Milk Fat

Fortnightly fat content (%) in milk of cows fed with or without protected fat and protein are presented in table 15. From the table it is seen that the milk fat content (Table 15) ranged from 3.77 to 3.95, 3.53 to 4.14, 3.89 to 4.23 and 4.36 to 4.79 per cent in T_0, T_1, T_2 and T_3 respectively in different fortnights. The overall average milk fat per cent was significantly higher ($P<0.05$) in group T_3 (4.52) than T_2 (4.03), T_0 (3.84) and T_1 (3.81). However, T_0, T_1, and T_2 were at par to each other. The higher fat per cent in T_3 treatment might be due to feeding of protected fat and protein in combination, which might have served as precursor for synthesis of milk fat. In the present study, the increase in milk fat per cent in the rumen protected fat and protein supplemented group (T_3) might be due to several reasons. Firstly, availability of more fatty acids (saturated and unsaturated) for absorption in intestine due to protection of fat and these fatty acids might have been directly incorporated in milk fat after absorption from intestine, leading to increase in milk fat. Secondly, due to formaldehyde treatment to groundnut cake, inefficiency in energy utilization was minimized and might have resulted in increased milk fat synthesis. Thirdly, the quality and quantity of amino acids available at intestines for the absorption were perhaps enhanced due to the treatment of cakes, because formaldehyde treatment of cakes might have protected these amino acids from ruminal degradation. Relatively higher supply of amino acids (limiting as well as non-limiting) at tissue level might have enhanced the efficiency of energy and protein utilization, and thus contributed indirectly to the increased milk fat synthesis. Among all components of milk, fat content is most sensitive to the dietary changes. Moreover, about 50 per cent of fat found in milk is synthesized in mammary gland from acetate and butyrate, while 40 to 45 per cent from the dietary source and less than 10 percent are derived from the mobilization of adipose tissue (Palmquist and Jenkins, 1980). So, supplemental fat source can increase milk fat of dairy cows. On supplementation of bypass fat to lactating animals, milk fat percentage is either increased [(West and Hill (1990); Sklan *et al.*, 1991; Sarwar *et al.*(2003); Barley and Baghul (2009); Thakur and Shelke (2010); Sirohi *et al.*, 2010; Zhang *et al.* (2011); Sontakke *et al.* (2014); Kirovski *et al.* (2015)] or decreased [(Chouinard *et al.*, 1998)] or not altered [(Furguson *et al.* (1990); Lounglawan *et al.* (2007); Puroshothaman *et al.* (2008); Naik *et al.* (2009b); Tyagi *et al.* (2009a); Ranjan *et al.* (2012); Singh *et al.* (2014) and Yadav *et al.* (2015)]. Also on supplementation of bypass protein to lactating animals, milk fat percentage is either increased [(Gulati *et al.* (2002); Chatterjee and Walli (2003); Garg *et al.* (2003b); Sampath *et al.* (2005); Kumar *et al.* (2005); Garg *et al.* (2007); Mishra *et al.* (2008); Chandrashekharan *et al.*

(2008); Dosky *et al.* (2012); Sherasia *et al.* (2012); Amrutkar *et al.* (2014) and Quisi and Titi (2014)] or not altered [(Trinacly *et al.* (2006); Sirohi *et al.* (2013)].

Table 15: Fortnightly fat content (%) in milk of cows fed with or without protected fat and protein

Treat/days	15	30	45	60	75	90	Mean
T_0	3.77	3.88	3.81	3.82	3.95	3.82	**3.84[b]**
T_1	4.14	3.90	3.66	3.83	3.77	3.53	**3.81[b]**
T_2	3.99	3.89	3.93	4.14	4.03	4.23	**4.03[b]**
T_3	4.47	4.66	4.79	4.43	4.36	4.38	**4.52[a]**
Mean	**4.09**	**4.08**	**4.05**	**4.05**	**4.02**	**3.99**	**4.05**
	S	T	SxT	CV%			
SEm	0.100	0.082	0.201				
CD5%	NS	0.227	NS	12.879			

abc bearing different superscripts within row and coloum different significantly (P<0.5)

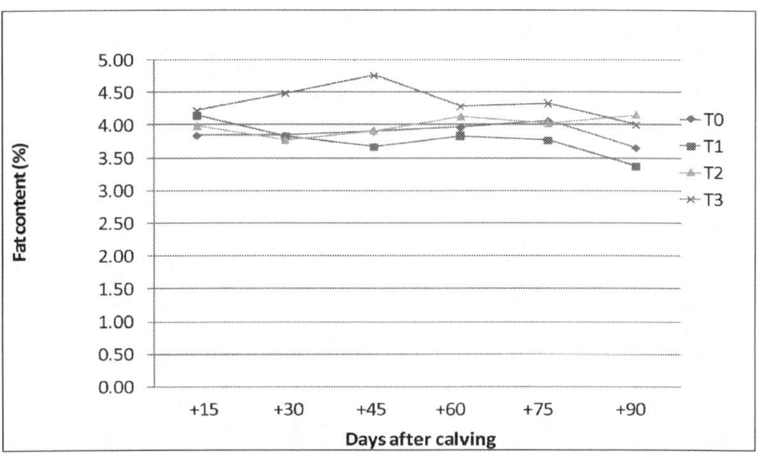

Fig. 6: Fortnightly fat content (%) in milk of cows fed with or without protected fat and protein

However, on supplementation of bypass fat and protein, milk fat percentage is either increased [(Shelke *et al.* (2012b); Dhulipalla *et al.* (2013); Vahora *et al.* (2013); Grewal *et al.* (2014)] or not altered [(Strusinska *et al.* (2006 and Nam *et al.* (2014)].

Table 16: Fortnightly fat yield (Kg/d) in cattle fed with or without protected fat and protein

Treat	15	30	45	60	75	90	Mean
T_0	0.31	0.42	0.42	0.41	0.40	0.36	**0.39**[b]
T_1	0.41	0.44	0.47	0.47	0.45	0.42	**0.44**[b]
T_2	0.36	0.42	0.49	0.49	0.51	0.52	**0.46**[b]
T_3	0.41	0.55	0.63	0.58	0.58	0.49	**0.54**[a]
Mean	**0.37**[b]	**0.46**[a]	**0.50**[a]	**0.49**[a]	**0.49**[a]	**0.45**[a]	**0.46**
	S	T	SxT	CV%			
SEm	0.027	0.022	0.053				
CD5%	0.074	0.060	NS	36.341			

abc bearing different superscripts within row and coloum different significantly (P<0.5)

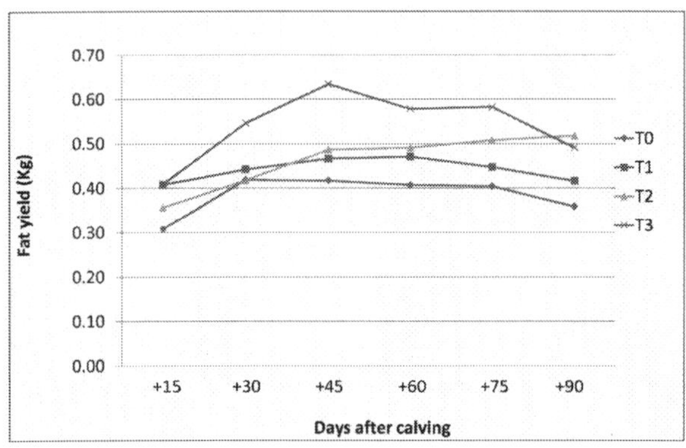

Fig. 7: Fortnightly fat yield (Kg/d) in cattle fed with or without protected fat and protein

Fortnightly fat yield (kg/day) in milk of cows fed with or without protected fat and protein are presented in table 16.Total fat yield kg/ day was significantly higher in T_3 (0.56) than T_2 (0.46), T_1 (0.44) and T_0 (0.39) which were at par. These results are in agreement with that of Naik *et al.* (2009b) who reported addition of bypass fat in diet generally increases the total milk fat yield due to increase in the milk production. Similar results were noticed by Puroshothaman *et al.* (2008) by feeding bypass fat; Kumar *et al.* (2005) and Dosky *et al.* (2012) by feeding bypass protein and Shelke et al (2011) and Nam *et al.* (2014) by feeding bypass fat and protein.

4.9.2.2 Protein

Fortnightly protein (%) in milk of cows fed with or without protected fat and protein are presented in table 17. The protein content in milk ranged from 3.01 to 3.18, 2.90 to 3.07, 2.95 to 3.06 and 2.96 to 3.12 per cent in T_0, T_1, T_2 and T_3, respectively in different fortnights. The average protein content in T_0, T_1, T_2, and T_3 were 3.06, 2.96, 2.97 and 3.03 per cent respectively. The protein content in present study was observed significantly higher in T_0 followed by T_3 and almost same in T_2 and T_1 treatment group. There was undesirable decreasing trend in milk protein content with advancement of period, however the interaction effect is non significant, indicating that, whatever extra protein and fat was made available through protection that had been utilized for milk production purpose. This is seen from earlier table that the milk yield was more in all treatment groups (Table 14). Similar observations were reported in protein and fat yield too (Table 16 and 18). Generally, supplementation of bypass fat (Ca-LCFA) has negative effect on the milk protein percentage; an overall effect of -0.12 percentage unit (Chouinard *et al.*, 1998). Depressed milk protein percentage is related to the dilution of milk protein as higher milk volume synthesized is not synchronized with uptake of amino acids by the mammary gland (DePeters and Cantt, 1992). Further, dietary fat impairs amino acids transport to mammary gland and decreases milk protein synthesis by inducing insulin resistance (Palmquist and Moser, 1981). If protein or amino acids are inadequate, lipoprotein synthesis may be decreased. Dietary protein should be increased, when fat supplemented diets are fed to animals (Palmquist and Moser, 1981). However, non significant effect reported by Sarwar *et al.* (2003); Louglawan *et al.* (2006); Purushothaman *et al.* (2008); Naik *et al.* (2009b); Tyagi *et al.*, 2009a; Thakur and Shelke, 2010; Sirohi *et al.*, 2010); Zhang *et al.* (2011); Ranjan *et al.* (2012); Singh *et al.* (2014); Sontakke *et al.* (2014); Kirovski *et al.* (2015) and Yadav *et al.* (2015) or increased (Wadhwa *et al.*, 2012) or decreased (West and Hill (1990); Chouinard *et al.* 1998); De peters and Cantt (1992) in milk protein percentage are also available.

Also on supplementation of bypass protein to lactating animals, milk protein percentage was either increased as reported by Canale *et al.* (1990); Cantt *et al.* (1991); Garg *et al.* (2002); Gulati *et al.* (2002); Chatterjee and Walli (2003); Kumar *et al.* (2005); Garg *et al.* (2007); Mishra *et al.* (2006); Dosky *et al.* (2012) or not altered according to Garg *et al.* (2003); Sirohi *et al.* (2013) and Amrutkar *et al.* (2014).

Table 17: Fortnightly protein (%) in milk of cow fed with or without protected fat and protein

Treat/days	15	30	45	60	75	90	Mean
T_0	3.18	3.01	3.04	3.05	3.01	3.07	**3.06^a**
T_1	3.07	2.90	2.92	2.93	2.97	2.97	**2.96^c**
T_2	3.06	2.95	2.96	2.97	2.95	2.95	**2.97$_{bc}$**
T_3	3.18	3.06	2.99	2.99	2.96	2.98	**3.03$_{ab}$**
Mean	**3.12^a**	**2.98^b**	**2.97^b**	**2.98^b**	**2.97^b**	**2.99^b**	**3.00**
	S	T	SxT	CV%			
SEm	0.027	0.022	0.055				
CD5%	0.076	0.062	NS	4.389			

abc bearing different superscripts within row and coloum different significantly (P<0.5)

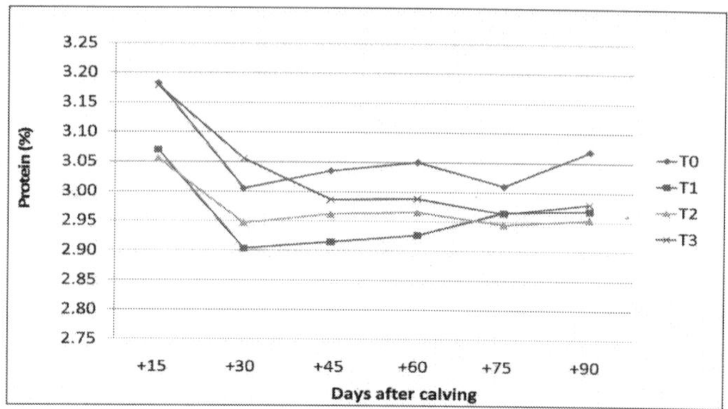

Fig. 8: Fortnightly protein (%) in cattle fed with or without protected fat and protein

However, on supplementation of bypass fat and protein, milk protein percentage is either increased (Dhulipalla *et al.* (2013) or not altered [(Strusinska *et al.* (2006); Shelke and Thakur (2011); Grewal *et al.* (2014); Nam *et al.* (2014)].

Fortnightly milk protein yield (kg) in cattle fed with or without protected fat and protein are presented in table 18. From the table it is revealed that total protein yield kg/day significantly higher (P<0.05) in T_3 (0.38) followed by T_1 (0.35), T_2(0.34) than T_0(0.30). The treatment and period effect was found to be significant on average protein yield of the animals. Trinacly *et al.* (2006) reported increased protein yield by feeding rumen protected protein supplemented with amino acids lysine, methionine and histidine. Naik *et al.,* 2009b reported the total milk protein yield is increased due to the increase in milk production. Dosky *et al.* (2012) observed increased significantly protein yield by feeding protected soyabean meal (SBM) to Meriz does. Nam *et al.* (2014) reported significantly increased milk protein yield in supplemented rumen protected amino acids and rumen protected fat groups.

Table 18: Fortnightly protein yield (kg/d) in cattle fed with or without protected fat and protein

Treat/days	15	30	45	60	75	90	Mean
T_0	0.25	0.32	0.32	0.31	0.30	0.29	**0.30**$_b$
T_1	0.30	0.33	0.37	0.37	0.35	0.36	**0.35**$_a$
T_2	0.27	0.32	0.36	0.35	0.37	0.37	**0.34**$_a$
T_3	0.32	0.38	0.39	0.40	0.40	0.36	**0.38**$_a$
Mean	**0.28**$_b$	**0.34**$_a$	**0.36**$_a$	**0.36**$_a$	**0.35**$_a$	**0.35**$_a$	**0.34**
	S	T	SxT	CV%			
SEm	0.016	0.013	0.031				
CD5%	0.043	0.035	NS	25.95			

abc bearing different superscripts within row and coloum different significantly (P<0.5)

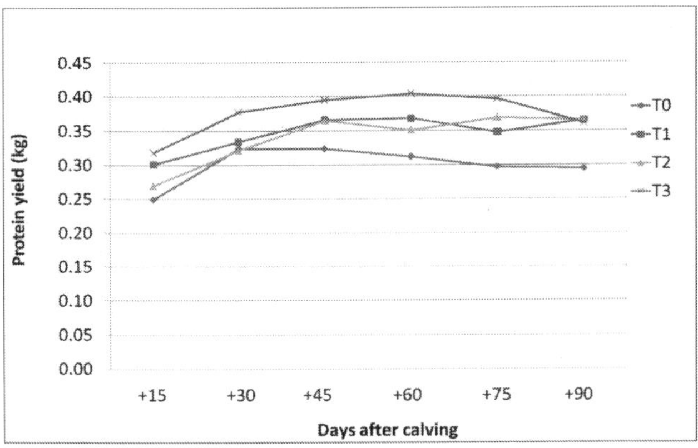

Fig. 9: Fortnightly protein yield (kg) in cattle fed with or without protected fat and protein

4.9.2.3 Fat corrected milk yield

Fortnightly average fat corrected milk yield (4% FCM) (kg/day) in cattle fed with or without protected fat and protein are presented in table 19. Average 4% FCM yield ranged from 7.66 to 10.66 in T_0, 10.00 to 12.10 in T_1, 8.83 to 12.82 in T_2 and 10.57 to 14.82 kg/d in T_3 in different fortnights supplementation periods. The average daily 4% FCM yield were 9.66, 11.40, 11.56 and 13.37 kg/d in T_0, T_1, T_2 and T_3, respectively. Group T_1, T_2 and T_3 had 15.26, 16.43 and 27.74% higher FCM yield over T_0 group. Decline in milk yield to progress of lactation was much slower in treatment groups. The treatment and period significantly affected average FCM yield of the animals. Significantly higher FCM was observed in T_3 over all the treatment groups but T_1 and T_2 were at par whereas; T_0 was significantly lower among treatment group.

Table 19: Fortnightly average FCM yield (kg/d) in cattle fed with or without protected fat and protein

Treat/days	**15**	**30**	**45**	**60**	**75**	**90**	**Mean**	**%↓↑**
T_0	07.66	10.66	10.42	10.01	09.84	09.36	**09.66^c**	
T_1	10.00	11.34	12.04	12.10	11.42	11.50	**11.40^b**	**15.26**
T_2	08.83	10.77	12.22	12.09	12.61	12.82	**11.56^b**	**16.43**
T_3	10.57	13.48	14.82	14.39	14.15	12.81	**13.37^a**	**27.74**
Mean	**09.26^b**	**11.56^a**	**12.38^a**	**12.15^a**	**12.00^a**	**11.62^a**	**11.50**	
	S	T	SxT	CV%				
SEm	0.558	0.455	1.116					
CD5%	1.546	1.262	NS	29.201				

abc bearing different superscripts within row and coloum different significantly (P<0.5)

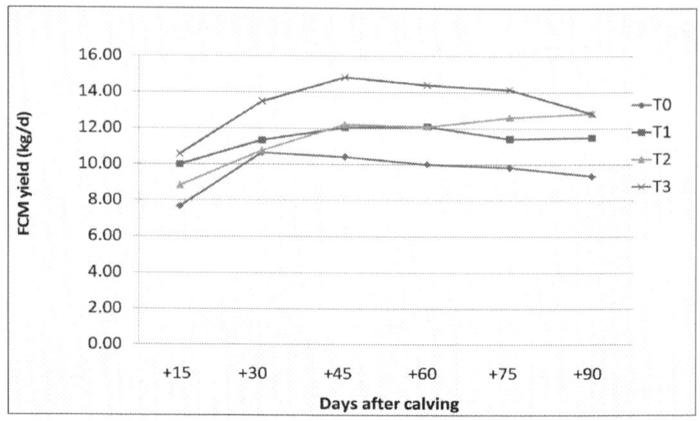

Fig. 10: Fortnightly average FCM yield (kg/d) in cattle fed with or without protected fat and protein

Schneider *et al.* (1988); Sklan *et al.* (1991); Schauff and Clark (1992); Sarwar *et al.* (2003); Purushothaman *et al.* (2008); Sirohi *et al.*, 2010); Ranjan *et al.* (2012); Sontakke *et al.* (2014) Ramteke *et al.* (2014) and Yadav *et al.* (2015) reported significantly increased FCM yield by feeding bypass fat. Whereas, Sampath *et al.* (1997); Yadav and Chaudhary (2004); Garg *et al.* (2005); Vahora *et al.* (2012); Sirohi *et al.* (2013) and Amrutkar *et al.* (2014) reported significantly increased FCM yield by feeding bypass protein. Also Strusinska *et al.* (2006); Shelke *et al.* (2012b); Dhulipalla *et al.* (2013); Vahora *et al.* (2013) and Nam *et al.* (2014) reported significantly increased FCM yield by feeding bypass fat and protein.

4.9.2.4 Energy corrected milk yield (ECM)

Fortnightly energy corrected milk yield (ECM) (kg/day) in cattle fed with or without protected fat and protein are presented in table 20. From the table it is revealed that the daily energy corrected milk yield during supplementation period was 9.54, 11.08, 11.20 and 12.68 kg/d in group T_0, T_1, T_2 and T_3, respectively which was 24.61 % significantly higher in T_3, 14.82% higher in T_2 and 13.89% in T_1 than T_0 whereas treatment T_2 and T_1 were at parwith each other and significantly lower ECM found in T_0. Ramteke *et al.* (2014) reported significantly increased ECM yield by feeding bypass fat and Amrutkar *et al.* (2014) by feeding bypass protein.

Table 20: Fortnightly average ECM yield (kg/d) in cattle fed with or without protected fat and protein

Treat	15	30	45	60	75	90	Mean
T_0	07.70	10.38	10.33	10.01	09.74	09.09	**09.54[c]**
T_1	09.80	10.92	11.75	11.82	11.17	11.05	**11.08[b]**
T_2	08.66	10.36	11.90	11.73	12.25	12.31	**11.20[b]**
T_3	10.03	12.75	14.16	13.66	13.61	11.89	**12.68[a]**
Mean	**09.05[b]**	**11.10[a]**	**12.03[a]**	**11.80[a]**	**11.69[a]**	**11.08[a]**	**11.13**
	S	T	SxT	CV%			
SEm	0.550	0.449	1.100				
CD5%	1.525	1.245	NS	29.651			

abc bearing different superscripts within row and coloum different significantly (P<0.5)

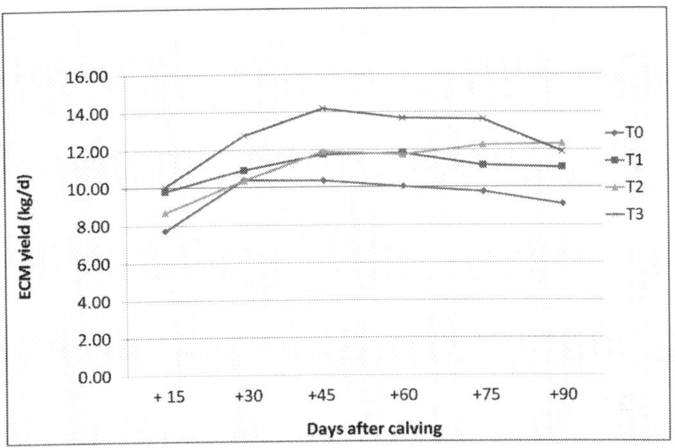

Fig. 11: Fortnightly average ECM yield (kg/d) in cattle fed with or without protected fat and protein

4.9.2.5 Milk lactose

Fortnightly lactose content (%) and lactose yield (kg/day) in milk of cattle fed with or without protected fat and protein are presented in table 21 and 22. The milk lactose ranged from 4.86 to 5.04, 4.67 to 4.92, 4.71 to 4.93 and 4.71 to 5.10 per cent in T_0, T_1, T_2 and T_3 respectively. The mean values were 4.93, 4.76, 4.77 and 4.80 per cent in T_0, T_1, T_2, and T_3 respectively. The lactose content was significantly higher in T_0 over T_1, T_2 and T_3, respectively. Furguson *et al.* (1989); Louglawan *et al.* (2006); Purushothaman *et al.* (2008); Naik *et al.* (2009b); Sirohi *et al.*, 2010); Zhang *et al.* (2011); Ranjan *et al.* (2012); Singh *et al.* (2014); Sontakke *et al.* (2014) and Yadav *et al.* (2015) reported no change in lactose percentage by feeding bypass fat and increased lactose reported by Wadhwa *et al.* (2012). Also by feeding bypass protein, Trinacly *et al.* (2006);

Sirohi *et al.* (2013) and Amrutkar *et al.* (2014) reported no change in lactose percentage and Kumar *et al.* (2005) and Sampath *et al.* (2005) reported increasing trend.

Table 21: Fortnightly milk lactose (%) in cattle fed with or without protected fat and protein

Treat/days	15	30	45	60	75	90	Mean
T_0	5.04	4.86	4.91	4.88	4.88	4.99	**4.93^{a}**
T_1	4.92	4.71	4.69	4.67	4.77	4.82	**4.76^{b}**
T_2	4.93	4.72	4.78	4.71	4.77	4.73	**4.77^{b}**
T_3	5.10	4.83	4.64	4.74	4.71	4.77	**4.80^{b}**
Mean	**5.00^{b}**	**4.78^{a}**	**4.75^{a}**	**4.75^{a}**	**4.78^{a}**	**4.83^{a}**	**4.81**
	S	T	SxT	CV%			
SEm	0.037	0.030	0.073				
CD5%	0.101	0.083	NS	3.590			

abc bearing different superscripts within row and coloum different significantly (P<0.5)

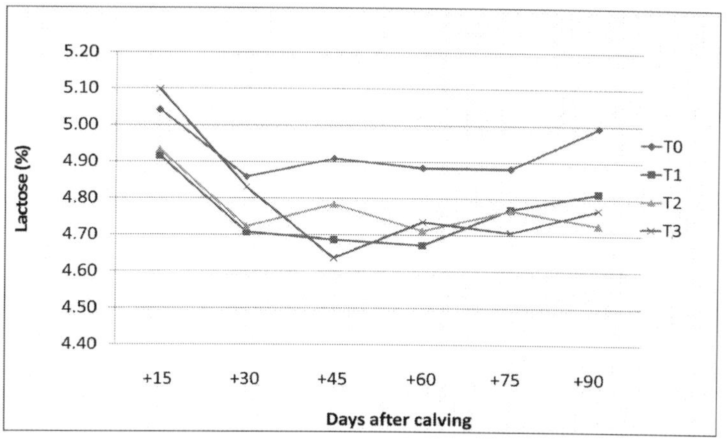

Fig. 12: Fortnightly lactose (%) in cattle fed with or without protected fat and protein

However, Strusinska *et al.* (2006) and Shelke *et al.* (2012b) reported no change and Dhulipalla *et al.* (2013) reported increase in lactose percentage by feeding bypass fat and protein.

The total lactose yield (Table 22) was significantly higher in T_3 (0.59) followed by T_1 (0.56) and T_2 (0.55) than control T_0 (0.48). These results were in agreement with Naik *et al.* (2009b) who reported that total milk lactose yield increased due to the increase in milk production by feeding bypass fat.

Table 22: Fortnightly lactose yield (kg/d) in cattle fed with or without protected fat and protein

Treat	15	30	45	60	75	90	Mean
T_0	0.39	0.52	0.52	0.50	0.48	0.48	**0.48[b]**
T_1	0.48	0.54	0.59	0.59	0.56	0.59	**0.56[a]**
T_2	0.43	0.51	0.59	0.56	0.60	0.58	**0.55[a]**
T_3	0.51	0.60	0.61	0.64	0.63	0.58	**0.59[a]**
Mean	**0.46[b]**	**0.54[a]**	**0.58[a]**	**0.57[a]**	**0.57[a]**	**0.56[a]**	**0.54**
	S	T	SxT	CV%			
SEm	0.024	0.020	0.048				
CD5%	0.067	0.054	NS	24.741			

abc bearing different superscripts within row and coloum different significantly (P<0.5)

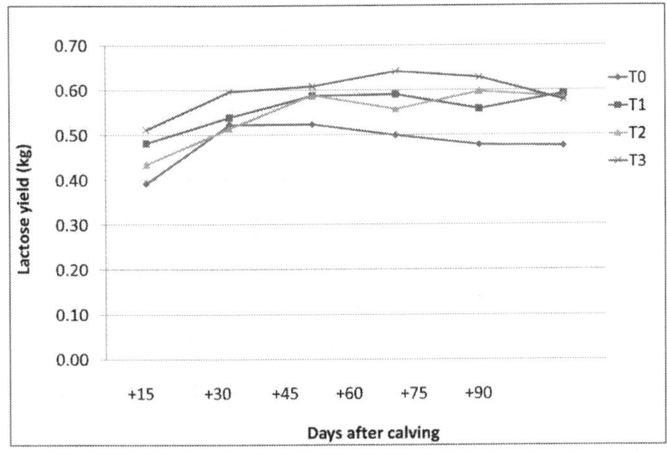

Fig. 13: Fortnightly lactose yield (kg) in cattle fed with or without protected fat and protein

4.9.2.6 Total solids

Fortnightly total solid content (%) and total solid yield (kg/day) of cows fed with or without protected fat and protein are presented in table 23 and 24, respectively. From the table it is revealed that the total solids ranged from 12.53 to 12.80, 12.05 to 12.94, 12.33 to 12.73 and 12.82 to 13.67 in T_0, T_1, T_2, and T_3, respectively. The mean values of total solids were 12.60, 12.33, 12.56 and 13.16 in T_0, T_1, T_2 and T_3, respectively, which was significantly higher in T_3 treatment groups over T_0, T_1, and T_2 treatment group. Similar results were reported by Furguson *et al.* (1989); Louglawan *et al.* (2006); Purushothaman *et al.* (2008); Naik *et al.* (2009b); Ranjan *et al.* (2012); Sontakke *et al.* (2014) and Ramteke *et al.* ((2014) in total solids content by feeding bypass fat. Trinacly *et al.* (2006); Sirohi *et al.* (2013) reported no change in total solids content by feeding rumen protected protein supplement with amino acid and bypass protein whereas Kumar *et al.* (2005); Sampath *et al.* (2005); Mishra *et al.* (2006);

Chandrashekharan *et al.*(2008) and Amrutkar *et al.* (2014) reported increasing trend in total solids content by feeding bypass protein. However, Strusinska *et al.* (2006) and Shelke *et al.* (2012b) reported no change and Dhulipalla *et al.* (2013) and Vahora *et al.* (2013) reported significantly increase in total solids content by feeding bypass fat and protein.

Table 23: Fortnightly total solid (%) content in milk of cow fed with or without protected fat and protein

Treat/days	15	30	45	60	75	90	Mean
T_0	12.80	12.53	12.54	12.55	12.56	12.61	**12.60^b**
T_1	12.94	12.40	12.05	12.22	12.28	12.09	**12.33^b**
T_2	12.73	12.33	12.45	12.61	12.57	12.69	**12.56^b**
T_3	13.67	13.49	13.19	12.90	12.82	12.93	**13.16^a**
Mean	**13.03**	**12.69**	**12.56**	**12.57**	**12.56**	**12.58**	**12.66**
	S	T	SxT	CV%			
SEm	0.139	0.113	0.278				
CD5%	NS	0.314	NS	5.391			

abc bearing different superscripts within row and coloum different significantly (P<0.5)

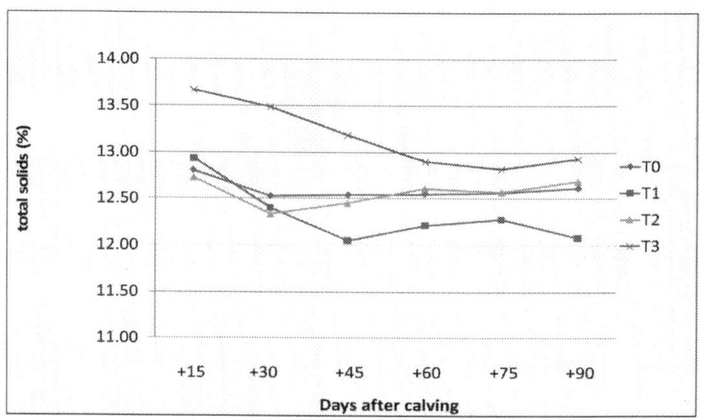

Fig. 14: Fortnightly total solid (%) content in milk of cow fed with or without protected fat and protein

Table 24: Fortnightly total solids yield (kg/d) in cattle fed with or without protected fat and protein

Treat	15	30	45	60	75	90	Mean
T_0	1.00	1.35	1.34	1.29	1.24	1.21	**1.24^c**
T_1	1.27	1.42	1.52	1.53	1.44	1.49	**1.45^b**
T_2	1.13	1.35	1.54	1.49	1.58	1.57	**1.44^b**
T_3	1.36	1.66	1.74	1.75	1.72	1.57	**1.63^a**
Mean	**1.19^b**	**1.45^a**	**1.53^a**	**1.51^a**	**1.49^a**	**1.46^a**	**1.44**
	S	T	SxT	CV%			
SEm	0.067	0.055	0.134				
CD5%	0.186	0.152	NS	27.066			

abc bearing different superscripts within row and coloum different significantly (P<0.5)

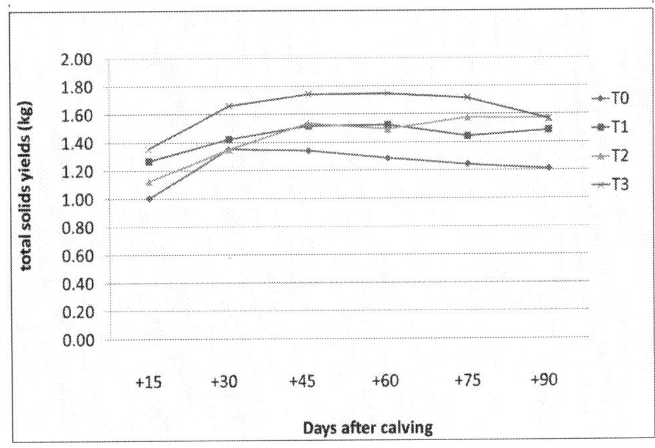

Fig. 15: Fortnightly total solids yield (kg) in cattle fed with or without protected fat and protein

Total yield of total solid kg/day was significantly higher in T_3 (1.63) than T_1 (1.45) and T_2 (1.44) but at par with each other and significantly lower in T_0 (1.24). Naik *et al.* 2009b reported that total solid contents are not influenced by the supplemental bypass fat.

4.9.2.7 Solid not fat (SNF)

Fortnightly Solid Not Fat (SNF) content (%) and SNF yield (kg/day) of cows fed with or without protected fat and protein are presented in table 25 and 26, respectively. The SNF content (Table 25) ranged from 8.62 to 9.03, 8.38 to 8.79, 8.45 to 8.74 and 8.40 to 9.20 per cent in T_0, T_1, T_2 and T_3, respectively. The average values of SNF were 8.76, 8.52, 8.53 and 8.65 in T_0, T_1, T_2, and T_3, respectively which was significantly higher in T_0 and T_3 over T_1 and T_2.

Table 25: Fortnightly SNF (%) content in milk of cow fed with or without protected fat and protein

Treat/days	**15**	**30**	**45**	**60**	**75**	**90**	**Mean**
T_0	9.03	8.65	8.73	8.73	8.62	8.80	**8.76[a]**
T_1	8.79	8.50	8.40	8.38	8.52	8.55	**8.52[b]**
T_2	8.74	8.45	8.52	8.47	8.54	8.46	**8.53[b]**
T_3	9.20	8.83	8.40	8.48	8.46	8.55	**8.65[ab]**
Mean	**8.94[a]**	**8.61[b]**	**8.51[b]**	**8.52[b]**	**8.53[b]**	**8.59[b]**	**8.62**
	S	T	SxT	CV%			
SEm	0.071	0.058	0.142				
CD5%	0.197	0.160	NS	3.949			

abc bearing different superscripts within row and coloum different significantly (P<0.5)

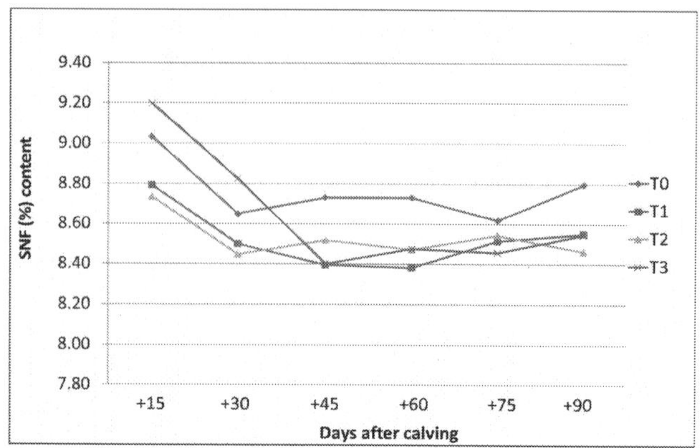

Fig. 16: Fortnightly SNF (%) content in milk of cow fed with or without protected fat and protein

The SNF content of milk was either not altered [Furguson *et al.* (1989); Louglawa *et al.* (2006); Purushothaman *et al.* (2008); Naik *et al.* (2009b); Ranjan *et al.* (2012); Sontakke *et al.* (2014) and Ramteke *et al.* ((2014)] or increased [(Wadhwa *et al.*, 2012) and Yadav *et al.* (2015)] or decreased [Ramtake *et al.* (2014)] by feeding bypass fat. Kumar *et al.* (2005); Sirohi *et al.* (2013) and Amrutkar *et al.* (2014 reported no difference in SNF content by feeding bypass protein whereas Mishra *et al.* (2006); Chandrashekharan *et al.* (2008) reported significantly increased SNF content by feeding bypass protein. However, Shelke *et al.* (2011) reported no change in SNF content by feeding bypass fat and protein.

Table 26: Fortnightly SNF yield (kg/d) in cattle fed with or without protected fat and protein

Treat	15	30	45	60	75	90	Mean
T_0	0.69	0.93	0.93	0.89	0.85	0.84	**0.86**[b]
T_1	0.86	0.97	1.05	1.05	1.00	1.05	**1.00**[a]
T_2	0.77	0.92	1.05	1.00	1.07	1.05	**0.98**[a]
T_3	0.92	1.09	1.10	1.15	1.13	1.03	**1.07**[a]
Mean	**0.81**[a]	**0.98**[b]	**1.03**[b]	**1.02**[b]	**1.01**[b]	**0.99**[b]	**0.98**
	S	T	SxT	CV%			
SEm	0.044	0.036	0.087				
CD5%	0.121	0.099	NS	25.339			

abc bearing different superscripts within row and coloum different significantly (P<0.5)

Total yield (Table 26) of SNF kg/day was significantly higher in T_3 (1.07) followed T_1 (1.00) and T_2 (0.98) over T_0 (0.86). Naik *et al.* (2009b) reported that the total SNF yield is increased due to the increase in milk production by feeding bypass fat.

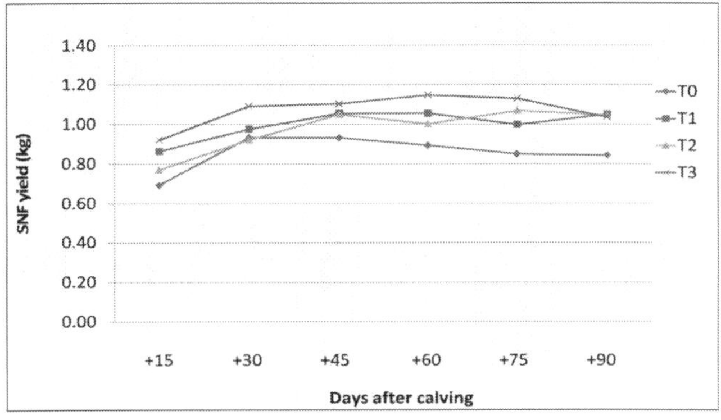

Fig. 17: Fortnightly SNF yield (kg) in cattle fed with or without protected fat and protein

4.9.2.8 Density

The milk density (Table 27) of experimental cow milk ranged from 29.33 to 30.20, 28.34 to 29.28, 28.75 to 29.78 and 28.21 to 29.08 g/cub in T_0, T_1, T_2 and T_3 respectively in different fortnights. The average density of milk of T_0, T_1, T_2, and T_3 were 29.88, 28.95, 29.10 and 28.63 per cent, respectively which did not differ from period (15) to period (90) days. However, it differs significantly between different treatments. Highest density (29.88) of milk was observed in T_0 (control) group. Where as, it was at par in T_1 and T_2 (28.95 and 29.10) and lowest in T_3 (28.63) which might be because of increased milk yield in treatment groups.

Table 27: Fortnightly density in cattle fed with or without protected fat and protein

Treat/days	15	30	45	60	75	90	Mean
T_0	30.20	29.88	30.12	29.76	29.33	29.98	**29.88^a**
T_1	29.21	28.65	28.97	28.34	29.28	29.27	**28.95^b**
T_2	29.70	28.78	29.39	28.90	29.09	28.75	**29.10^b**
T_3	28.71	28.48	28.21	28.83	28.50	29.08	**28.63^c**
Mean	**29.45**	**28.94**	**29.17**	**28.96**	**29.05**	**29.27**	**29.14**
	S	T	SxT	CV%			
SEm	0.230	0.188	0.460				
CD5%	NS	0.521	NS	3.762			

abc bearing different superscripts within row and coloum different significantly (P<0.5)

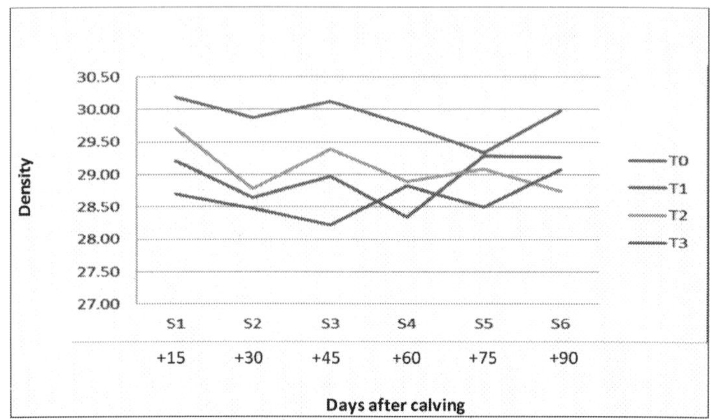

Fig. 18: Fortnightly Density of milk in cattle fed with or without protected fat and protein

4.9.2.9 Milk Temperature

The Fortnightly milk temperature of experimental cows milk is presented in Table 28 and ranged from 24.29 to 27.56, 23.33 to 25.94, 23.85 to 26.08 and 22.76 to 25.47 °C in T_0, T_1, T_2 and $T_{3,}$ respectively in different fortnights.

Table 28: Fortnightly temperature of milk (°C) of cattle fed with or without protected fat and protein

Treat/days	15	30	45	60	75	90	Mean
T_0	24.29	25.27	26.38	25.60	25.64	27.56	**25.79^b**
T_1	25.94	23.33	23.86	23.51	24.85	25.78	**24.54^a**
T_2	26.08	24.16	24.74	23.85	25.58	25.31	**24.95^a**
T_3	22.76	24.24	23.04	25.12	25.44	25.47	**24.35^a**
Mean	**24.77**	**24.25**	**24.50**	**24.52**	**25.38**	**26.03**	**24.91**
	S	T	SxT	CV%			
SEm	0.469	0.383	0.938				
CD5%	NS	1.061	NS	8.336			

abc bearing different superscripts within row and coloum different significantly (P<0.5)

The average milk temperature in T_0, T_1, T_2, and T_3 were 25.79, 24.54, 24.95 and 24.35 °C respectively which differ significantly higher in T_0 over treatment groups.

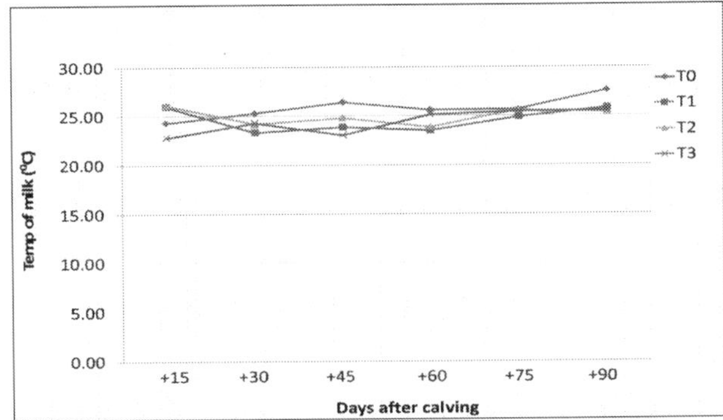

Fig. 19: Fortnightly temperature of milk (°C) of cattle fed with or without protected fat and protein

4.9.2.10 Freezing point

The freezing point of milk of experimental cows (Table 29) ranged from 0.57 to 0.58, 0.55 to 0.56, 0.56 to 0.58 and 0.55 to 0.56 0_c in T_0, T_1, T_2 and T_3 respectively in different fortnights. The average freezing point in T_0, T_1, T_2, and T_3 were 0.58, 0.56, 0.56 and 0.56 0_c, respectively which did not differ between the all groups.

Table 29: Fortnightly freezing point (°C) in cattle fed with or without protected fat and protein

Treat/days	15	30	45	60	75	90	Mean
T_0	0.58	0.57	0.58	0.58	0.57	0.58	**0.58^a**
T_1	0.56	0.56	0.55	0.56	0.56	0.56	**0.56^b**
T_2	0.58	0.56	0.56	0.55	0.57	0.56	**0.56^b**
T_3	0.56	0.56	0.55	0.56	0.56	0.56	**0.56^b**
Mean	**0.57**	**0.56**	**0.56**	**0.56**	**0.56**	**0.57**	**0.56**
	S	T	SxT	CV%			
SEm	0.005	0.004	0.010				
CD5%	NS	0.012	NS	4.322			

abc bearing different superscripts within row and coloum different significantly ($P<0.5$)

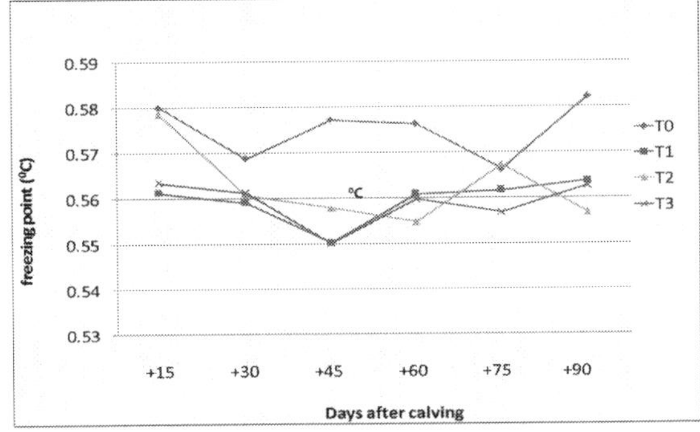

Fig. 20: Fortnightly freezing point (°C) in cattle fed with or without protected fat and protein

Table 30: Fortnightly mineral matter (%) in cattle fed with or without protected fat and protein

Treat/days	15	30	45	60	75	90	Mean
T_0	0.79	0.78	0.80	0.79	0.78	0.79	**0.79**
T_1	0.77	0.76	0.76	0.76	0.77	0.78	**0.77**
T_2	0.79	0.76	0.77	0.75	0.78	0.76	**0.77**
T_3	0.77	0.76	0.76	0.77	0.76	0.77	**0.76**
Mean	**0.78**	**0.76**	**0.77**	**0.76**	**0.77**	**0.78**	**0.77**
	S	T	SxT	CV%			
SEm	0.007	0.005	0.013				
CD5%	NS	0.015	NS	4.084			

abc bearing different superscripts within row and coloum different significantly (P<0.5)

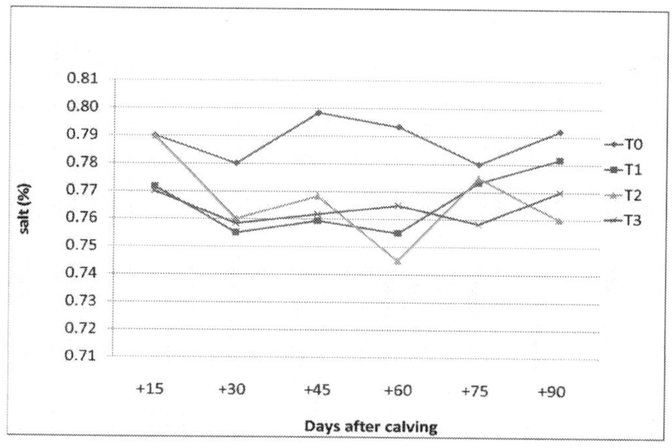

Fig. 21: Fortnightly mineral matter (%) in cattle fed with or without protected fat and protein

4.9.2.11 Mineral matter

The mineral matter (Table 30) ranged from 0.78 to 0.80, 0.76 to 0.78, 0.75 to 0.79 and 0.76 to o.77 per cent in T_0, T_1, T_2 and T_3, respectively in different fortnights. The average mineral content in T_0, T_1, T_2, and T_3 were 0.79, 0.77, 0.77 and 0.76 per cent, respectively which show significantly higher in control.

4.9.2.12 Fatty acid profile of milk of experimental cow

The fatty acid profiles of milk during experiment period of all groups are presented in Table 33.

Table 31: Saturated fatty acid profile (g/100g) of milk of cattle fed with or without protected fat and protein

	Saturated fat	T_0	T_1	T_2	T_3
C4:0	Butyric acid	1.95	1.94	1.80	1.75
C6:0	Caproic acid	2.70	2.75	2.55	2.45
C8:0	Caprylic acid	3.80	3.84	3.52	3.45
C10:0	Capric acid	3.18	3.33	3.28	3.38
C11:0	Undecanoic acid	0.00	0.00	0.00	0.00
C12:0	Lauric acid	4.55	4.30	4.20	4.01
C13:0	Tridecanoic acid	0.16	0.23	0.28	0.21
C14:0	Myristic acid	14.10	14.20	13.70	13.40
C15:0	Pentadecanoic acid	1.58	1.70	1.50	1.40
C16:0	Palmitic acid	22.40	22.64	22.55	22.60
C17:0	Heptadecanoic acid	0.00	0.00	0.00	0.00
C18:0	Stearic acid	12.80	12.90	12.57	12.38
C20:0	Arachidonic Acid	0.70	0.90	0.85	0.71
C22:0	Behenic Acid	0.00	0.00	0.00	0.00
C23:0	Tricosanoic Acid	0.00	0.00	0.00	0.00
C24:0	Lignoceric Acid	0.00	0.00	0.00	0.00
C21:0	Heneicosanoic acid	0.00	0.00	0.00	0.00
	Total	**68.32**	**67.73**	**65.88**	**65.24**

The table 31 shows that total saturated fatty acid content in T_0, T_1, T_2 and T_3 were 68.32, 67.73, 65.88 and 65.24 per cent, respectively.

Table 32: Unsaturated fatty acid profile (g/100g) of milk of cattle fed with or without protected fat and protein

	Fatty acid profile **Monounsaturated Fat**	**T_0**	**T_1**	**T_2**	**T_3**
C14:1	Myristoleic Acid	1.50	1.72	1.78	1.72
C15:1	Cis-10 Pentadecenoic	0.10	0.22	0.29	0.23
C16:1	Palmitoleic acid	0.00	0.00	0.00	0.00
C17:1	Cis 10Heptadecanoic acid	0.10	0.39	0.40	0.70
C18:1n	Oleic acid	19.54	19.85	20.28	20.63
C20:1	Cis-11 Eicosenoic acid	0.00	0.00	0.00	0.00
C22:1NG	Erucic acid	0.00	0.00	0.00	0.00
C24:1	Nervonic acid	0.00	0.00	0.00	0.00
Total		**21.24**	**22.18**	**22.75**	**23.28**
	Polyunsaturated Fat				
C18:2n6c	Conjugated Linoleic	2.80	2.90	3.20	3.10
C18:3n-3	Alpha-Linolenic acid	0.30	0.20	0.40	0.50
C18:3n-6	Gamma – linolenic acid	0.50	0.45	0.51	0.57
C20:2	Cis 11, 14- Eicosadienoic	0.00	0.00	0.00	0.00
C20:3n-3	Cis – 11,14,17- Eicosatrienoic	0.40	0.50	0.52	0.40
C20:3n-6	Cis–8,11,14–Eicosatrienoic acid	0.30	0.30	0.42	0.43
C20:4n-6	Arachidic acid	0.61	0.60	0.50	0.60

	Fatty acid profile Monounsaturated Fat	T_0	T_1	T_2	T_3
C20:5n-3	Cis–5,8,11,14,17, Eicosatgrienoic acid	0.00	0.00	0.00	0.00
C22:2	Cis – 13, 16 Docosadienoic acid	0.70	0.90	0.70	0.70
C22:6n-3	Cis-4,7,10,13,16,19 Docosadienoic acid	0.00	0.00	0.00	0.00
	Total	**5.61**	**5.85**	**6.25**	**6.30**
	Trans Fat				
C18:1n9t	Elaidic acid methyl ester	3.71	3.41	3.83	3.95
C18:2n6t	Linoledaidic Acid	1.12	0.83	1.29	3.95
	Total	**4.83**	**4.24**	**5.12**	**5.18**

The total unsaturated fatty acid content in T_0, T_1, T_2 and T_3, were 26.85, 28.03, 29.00 and 29.58 per cent, respectively. The monounsaturated fatty acid in T_0, T_1, T_2 and T_3, were 21.24, 22.18, 22.75 and 23.28 per cent, respectively.

Table 33: Fatty acid profile (%) of milk of cattle fed with or without protected fat and protein

Parameter	Type of Fatty acid (%)	T_0	T_1	T_2	T_3
1	**Saturated**	**68.32**	**67.73**	**65.88**	**65.24**
2	**Unsaturated**				
A	**Monounsaturated**	21.24	22.18	22.75	23.28
B	**Polyunsaturated**	5.61	5.85	6.25	6.30
3	**Total unsaturated**	**26.85**	**28.03**	**29.00**	**29.58**
4	**Other**	4.83	4.24	5.12	5.18
5	**Total fatty acid**	**100**	**100**	**100**	**100**

The poly unsaturated fatty acid in T_0, T_1, T_2 and T_3 are 5.61, 5.85, 6.25 and 6.30 per cent, respectively. Overall, there was increased concentration of mono, poly and unsaturated fatty acid by supplementing bypass fat and protein. These findings are in concurrence with the findings of several workers. Chouinard *et al.* (1997) reported that proportions of odd numbered FA (C 15:0, C17:0) of milk fat decrease because of ruminal bacteria which are major source of odd chain fatty acids of milk fat prefers to use preformed fatty acids. Mishra *et al.* (2004) reported decreases the proportions of short and medium chain saturated fatty acids (C6:0 to C16:0) of milk fat due to reduction in de novo FA synthesis in mammary gland and increase in proportions of LCFA (C18:1, C18:2, C18:3) due to increased uptake of preformed LCFA from blood by supplementation of Ca-LCFA in the diet of lactating cows. Lounglawan et al. (2007) reported rumen-bypass fat supplementation significantly ($P<0.05$) reduced C4:0 but increased C12:0 and C14:0 fatty acids in the cows' milk. Tyagi *et al.* (2009a) reported that supplementation of protected fat @ 2.5% DMI significantly increased the proportion of unsaturated FA and LCFA in milk fat. Purushothaman *et al.* (2008) reported that supplementation with calcium salts of palm oil fatty acid reduced the proportion of caproic, caprylic and capric acids and significantly (P<0.01) increased the concentration of palmitic, oleic, stearic, linoleic and linolenic acids in milk fat with increase in level of bypass fat supplementation. Tyagi *et al.* (2009) observed increased in the total USFA, LCFA and MUFA and decrease in the total SFA as percentage of the total fatty acids of milk due to supplementation of bypass fat in the diet of dairy cows. Thakur and Shelke (2010) reported increase in the proportion of unsaturated and long chain fatty acids in milk fat on supplementation of bypass fat prepared from soybean acid oil in lactating buffaloes. Tyagi *et al.* (2010) reported increased total long chain fatty acids (LCFA) and monounsaturated fatty acids (MUFA) by feeding bypass fat. Qureshi and Tawheed (2012) reported that SFA was significantly (P<0.05) decreased while MUFA and PUFA increased with the increasing supplementation of protected palm fats feeding on crossbred cows. It appears that hypercholestermic properties of the milk were reduced and cardio-protective properties were enhanced by feeding protected palm fats. Sontakke *et al.* (2014) observed significantly increase in USFA by feeding rice bran lyso-phospholipids (RBLP) and rumen protected fat (RPF) in crossbred lactating Karan Fries.

Trinacly *et al.* (2006) reported that proportion of UFA (both MUFA and PUFA) was increased and that of SFA in milk was decreased by feeding rumen protected protein supplemented with amino acids. Qaisi and Titi (2014) reported no differences in milk fatty acids composition by supplementing rumen protected methionine on goats.

Strusiñska *et al.* (2006) resulted in a significant increase in the concentrations of unsaturated fatty acids (especially C18:1 and C18:2) and hypocholesterolaemic acids (DFA) in milk by feeding fat and protein supplementation. Garg *et al.* (2012) reported improvement in long chain fatty acids (LCFA; C16:0 to C20:0) and mono-unsaturated fatty acids (MUFA; C14:1, C16:1 & C18:1) contents by feeding bypass fat with rumen protected choiline chloride. Also Shelke *et al.* (2012b) reported improvement in unsaturated fatty acid and decreased in saturated fatty acid by feeding bypass fat and protein in buffalo.

4.10 Effect of Protected Fat and Protein Supplementation on Reproductive Performance

4.10.1 Expulsion of fetal membranes and involution of uterus

4.10.1.1 Expulsion of placenta

Effect of protected fat and protein on reproductive performance is presented in table 34 and 35. From the table it is observed that, number of hours required for expulsion of placenta in T_0, T_1, T_2 and T_3 were 5.81, 4.11, 3.51 and 3.93 hrs, respectively, which was statistically (P<0.05) lower (2.30 hrs) in bypass fat supplemented group (T_2) than other groups. This might be due to better energy status of the animals, which is strongly associated with reproductive performance. The results of present study are in accordance with that of Tyagi *et al.* (2010) by feeding bypass fat in crossbred cows and Shelke *et al.* (2012a) who reported decrease in time required for expulsion of fetal membranes in bypass fat and protein (at 2.5% DMI) supplemented buffaloes.

4.10.1.2 Involution of uterus

Average numbers of days required for involution of uterus are presented in table 34. From the table it is revealed that number of days required for involution of uterus in T_0, T_1, T_2 and T_3 were 28.33, 27.50, 25.17 and 22.50 days, respectively which was statistically decreased (by 5.83 days) in bypass protein and fat group (T_3) over other groups. Prostaglands play an important role in restablishing estrous cycle both immediately after parturation and thereafter, until conception occures prostaglands F2? is responsible for uterine involution after parturation. This might be due to a positive effect of protected protein and fat supplementation on uterine functions. Reduction of incidence of RFM and less time taken for involution of uterus would facilitate the early and successful conception of the cow. The results of present study are in accordance with that of Tyagi *et al.* (2010) and Shelke *et al.* (2012a) who reported decrease in time required for involution of uterus in bypass fat (at 2.5% DMI) supplemented cows.

4.10.2 Reproduction related parameters

The variables recorded during post partum study are given in table 34.

4.10.2.1 First estrous observed after calving (Open Period)

The number of days required for appearance of first post partum oestrus in T_0, T_1, T_2 and T_3 were 87.83, 76.00, 66.00 and 61.83 days which was significantly (P< 0.05) reduced by 26 days in bypass protein and fat supplemented (T_3) group. The commencement of cyclicity was related with the process of involution of uterus, as the duration for uterine involution was reduced on bypass protein and fat supplementation, which may be responsible for relatively early onset of heat. The feeding of additional energy in the form of bypass fat reduces the cow's negative energy status so that cow returns to estrus earlier after calving and therefore conceives sooner. Palmquist and Jenkins (1980) reported that improved energy balance could result in an earlier return to postpartum ovarian cyclicity. The results of present study are in accordance with Sklan *et al.* (1991); Tyagi *et al.* (2010); Naik *et al.* (2009a); Gowda *et al.* (2013) by feeding bypass fat. Also Kaur and Arora (1995); Mishra *et al.* (2006) and Walli (2008) reported similar results by feeding bypass protein. Whereas, Shelke *et al.* (2012a) reported similar findings by feeding bypass fat and protein.

4.10.2.2 Service period and AI per conception

The observation on service period and AI per conception are recorded and depicted in table 34. The service period in T_0, T_1, T_2 and T_3 was 141.33, 130.67, and 82.00 and 99.67 days, respectively which was statistically reduced by 59.33 and 41.66 days in T_2 and T_3 groups than control. Similarly, number of AI per conceptions in T_0, T_1, T_2 and T_3 was 3.00, 2.50, 1.83 and 2.00, which was statistically lower in (T_3) bypass fat supplemented group. The service period and AI per conception was shorter in bypass fat group (T_2) than the other groups, indicating that lesser time was required for the cows in T_2 for conception. This might be due to slight increasing in the blood progesterone concentration in cows supplemented with bypass fat. Bypass fat increases concentration of circulating cholesterol, the precursor of progesterone, which then secreted by the corpus luteum helps in preparing the uterus for implantation of the embryo and further maintenance of pregnancy. The results of present study are in accordance with Tyagi *et al.* (2010); Naik *et al.* (2009); Gowda *et al.* (2013) by feeding bypass fat. Also Mishra *et al.* (2006) and Walli (2008) reported similar results by feeding bypass protein. Whereas, Shelke *et al.* (2012a) and Grewal *et al.* (2014) reported similar findings by feeding bypass fat and protein.

4.10.2.3 Conception rate

Highest conception rate (Table 34) was found in T_2 (66.67) treatment group followed by T_3 (54.55), T_1 (40.00) and lowest in T_0 (33.34). Similar results were reported by Ferguson *et al.* (1990) and Tyagi *et al.* (2010) on supplementing rumen inert fat in the diet of high yielding lactating dairy cows. Feeding Ca-LCFA increases pregnancy rate and reduces open days (Sklan *et al.*, 1991). Several hypotheses were suggested regarding role of the fatty acids on reproductive performance of dairy animals (Sklan *et al.*, 1994). These include (i) improved energy balance results in an earlier return to post-partum ovarian cycling; (ii) increase linoleic acid may provide increase PGF2α and stimulate return to ovarian cycling and improve follicular recruitment; and (iii) increase in progesterone secretion either from improved energy balance or from altered lipoprotein composition from dietary fat improves fertility. Due to bypass fat feeding, the average period for conception after calving was reduced (118 vs 92, days) in cows (Garg and Mehta, 1998). Naik *et al.* (2009b) reported that when bypass fat was included in diet of crossbred cows, the number of artificial inseminations required per conception was reduced (1.4 vs 1.2), indicating better reproductive performance of animals. However, changes in reproductive performance associated with fat supplementation are related to magnitude of the milk response of the fat supplementation (Scott *et al.*, 1995). The bypass fat supplementation during pre-partum period to advanced pregnant cows increased the calf weight (24.94 vs 27.95, kg), calving per cent (88.88 vs 100); decreased the incidences of still birth (1 vs 0), premature birth (1 vs 0) and retention of foetal membranes (4 vs 1) in high yielding crossbred cows (Tyagi *et al.*, 2009b). Gowda *et al.* (2013) also reported better reproductive performance in cows fed indigenously prepared bypass fat. Kaur and Arora (1995) reported increased conception rate by feeding bypass protein. The supplementation of protected fat and protein during early lactation improved the reproductive performance in Murrah buffaloes along with increase in the milk production and its persistency (Shelke *et al.*, 2012a) and Grewal *et al.* (2014).

Table 34: Effect of protected fat and protein on reproduction performance of cattle

Parameters	T_0	T_1	T_2	T_3	Mean	SE	CD 5%	CV%
EFM (hrs)	5.81[a]	4.11[b]	3.51[b]	3.93[b]	4.34	0.55	1.66	31.11
Involution of uterus (Days)	28.33[a]	27.50[a]	25.17[ab]	22.50[b]	25.88	01.13	3.41	10.71
Open period (Days)	87.83[a]	76.00[ab]	66.00[b]	61.83[b]	72.92	06.36	19.16	21.36
Service period (Days)	141.33[a]	130.67[ab]	82.00[c]	99.67[bc]	113.42	13.70	41.31	29.60
AI/Conception	03.00[a]	02.50[ab]	01.83[b]	02.00[b]	02.33	0.29	0.87	30.30
Conception rate (%)	33.34	40.00	66.67	54.55				

abc bearing different superscripts within row and coloum different significantly (P<0.5)

4.11 Effect of Protected Fat and Protein Supplementation on Blood Parameters

4.11.1 Albumin

The serum albumin concentration (mg/dl) during different fortnights is depicted in table 35 and varied from 3.05 to 3.43, 3.08 to 3.37, 3.07 to 3.36 and 3.07 to 3.37 mg/dl in T_0, T_1, T_2 and T_3 respectively. The overall average albumin concentration was 3.28, 3.21, 3.21 and 3.24 mg/dl in respective groups.

Table 35: Albumin concentration (mg/dl) of cattle fed with or without protected fat and protein

Treat/days	-30	-15	0	30	60	90	Mean
T_0	3.05	3.36	3.35	3.22	3.43	3.28	**3.28**
T_1	3.25	3.37	3.24	3.22	3.13	3.08	**3.21**
T_2	3.07	3.36	3.35	3.07	3.28	3.12	**3.21**
T_3	3.37	3.25	3.28	3.07	3.24	3.26	**3.24**
Mean	**3.18**	**3.34**	**3.30**	**3.14**	**3.27**	**3.18**	**3.24**
	S	T	SxT	CV%			
SEm	0.059	0.048	0.118				
CD5%	NS	NS	NS	8.796			

abc bearing different superscripts within row and coloum different significantly (P<0.5)

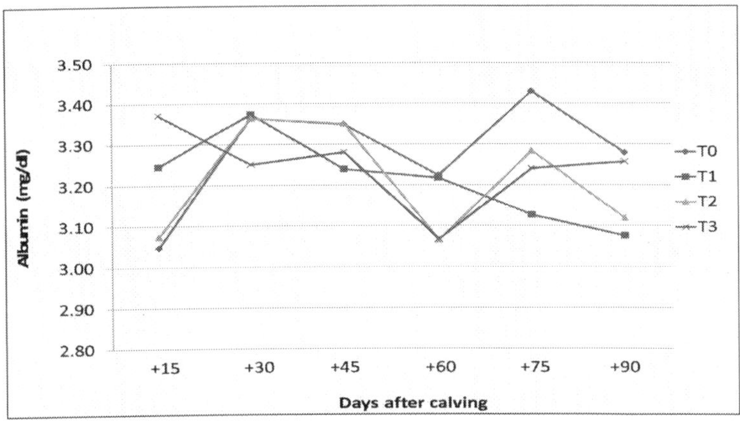

Fig. 22: Albumin concentration (mg/dl) of cattle fed with or without protected fat and protein

The albumin level remained within normal range and no difference was observed at any stage within treatment groups of experiment period. These findings are in accordance with Tyagi *et al.* (2009a); Wadhwa *et al.* (2012) and Kirovski *et al.* (2015) reported that bypass fat supplementation had no effect on the albumin. Also Moveliya *et al.* (2013) reported no effect and Sai *et al.* (2014) reported significantly increased trend of albumin level by feeding bypass protein. Whereas, Shelke *et al.* (2010) reported no effect on albumin level by feeding bypass fat and protein.

4.11.2 Globulin

The serum globulin mg/dl (Table 36) in different fortnights were 2.96 to 5.11 in T_0, 2.99 to 5.43 in T_1, 3.41 to 4.99 in T_2 and 3.34 to 4.73 in T_3. The overall mean was 4.06, 4.26, 3.99 and 3.97 in T_0, T_1, T_2 and T_3 respectively. From the table it is observed that, there was no statistical difference in all treatments. Similar observations were reported by Tyagi *et al.*, 2009a; Wadhwa *et al.*, 2012 on bypass fat supplementation. Also Moveliya *et al.* (2013) reported no effect and Sai *et al.* (2014) reported significantly increased trend of globulin level by feeding bypass protein. Whereas, Shelke *et al.* (2010) and Garg *et al.* (2012) reported no effect and Grewale *et al.* (2014) reported increased level of globulin by feeding bypass fat and protein.

Table 36: Globulin concentration (mg/dl) of cattle fed with or without protected fat and protein

Treat/days	-30	-15	0	30	60	90	Mean
T_0	2.96	3.45	3.25	5.11	4.90	4.71	**4.06**
T_1	3.10	3.47	2.99	5.43	5.38	5.18	**4.26**
T_2	3.84	3.49	3.41	4.99	4.23	3.95	**3.99**
T_3	3.69	3.87	3.34	4.45	4.73	3.75	**3.97**
Mean	**3.40^c**	**3.57^c**	**3.25^c**	**4.99^a**	**4.81^{ab}**	**4.40^b**	**4.07**
	S	T	SxT	CV%			
SEm	0.189	0.155	0.379				
CD5%	0.525	NS	NS	19.687			

abc bearing different superscripts within row and coloum different significantly (P<0.5)

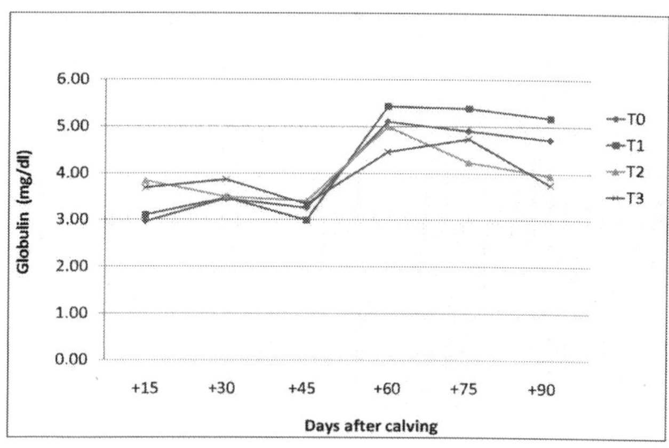

Fig. 23: Globulin concentration (mg/dl) of cattle fed with or without protected fat and protein

4.11.3 Blood urea nitrogen

Serum BUN mg/dl concentration presented in table 37 of different fortnight ranged between 15.86 to 17.83, 11.90 to 15.84, 14.72 to 16.99 and 12.64 to 16.29 mg/dl in T_0, T_1, T_2 and T_3 respectively. The average BUN values were 16.57, 13.69, 15.98 and 14.48 mg/dl in T_0, T_1, T_2 and T_3 respectively. The average BUN level was significantly lower (P<0.01) in T_1 and T_3 compared to T_0 and T_2 treatments groups. The treatment and period had significantly effect on BUN concentration of the animals. This may be due to protection of groundnut cake from microbial degradation which bypassed the rumen as intact protein. The higher value of blood urea level in T_0 and T_2 is indicative of less efficient utilization of dietary nitrogen for microbial protein synthesis due to higher ammonia level.

Feeding of rumen protected protein not only results in more supply of amino acids, but also saves energy wasted in urea synthesis. Blood urea concentration is an indicator of efficient protein balance (Campanile *et al.*, 1998; Dhali *et al.*, 2006) and is typically increased in cows deficient in energy.

Table 37: Blood urea nitrogen concentration (mg/dl) of cattle fed with or without protected fat and protein

Treat	-30	-15	0	30	60	90	Mean
T_0	17.83	16.42	16.14	16.49	15.86	16.69	**16.57^a**
T_1	14.85	15.84	13.09	11.90	13.18	13.28	**13.69^b**
T_2	16.99	16.85	15.23	14.72	16.00	16.07	**15.98^a**
T_3	16.29	15.36	14.83	12.64	13.83	13.90	**14.48^b**
Mean	**16.49^a**	**16.12^{ab}**	**14.82^c**	**13.94^c**	**14.72^c**	**14.98^{bc}**	**15.18**
	S	T	SxT	CV%			
SEm	0.461	0.376	0.922				
CD5%	1.277	1.043	NS	13.530			

abc bearing different superscripts within row and coloum different significantly (P<0.5)

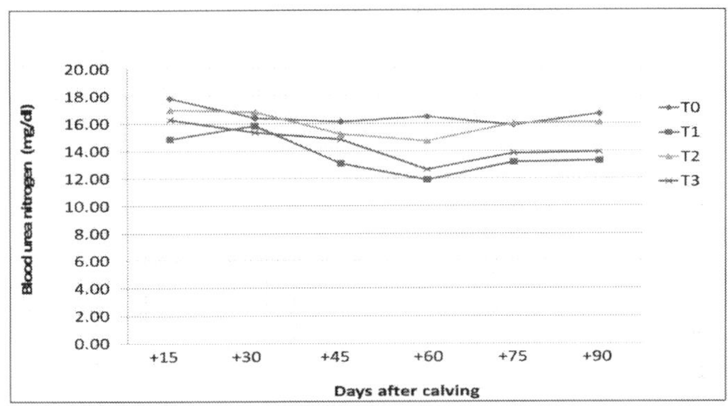

Fig. 24: Blood urea nitrogen concentration (mg/dl) of cattle fed with or without protected fat and protein

Yadav *et al.* (2015) reported significantly decresed BUN by feeding prilled fat to crossbred cows. Non-significant difference in BUN on supplementing bypass fat was reported by West and Hill (1990); Fahey *et al.* (2002); Tyagi *et al.* (2009a); Wadhwa *et al.* (2012) and Kirovski et al (2015). Similar results were reported by Tiwari and Yadav, (1994); Wankhede and Kalbande (2001); Sahoo and Walli, (2005); Torane *et al.* (2006); Yadav *et al.* (2006); Garg *et al.* (2007); Sherasia *et al.* (2012); Sai *et al.* (2014) and Moveliya *et al.* (2013) by feeding bypass protein supplements. Shelke *et al.* (2010) and Grewal *et al.* (2014) reported significantly decreased level of BUN by feeding bypass fat and protein. Whereas, Garg *et al.* (2012) reported non-significant difference in BUN on supplementing bypass fat and protein.

4.11.4 Blood cholesterol

The blood cholesterol concentration (mg/dl) is depicted in table 38 and ranged from 65.81 to 179.39, 76.48 to 204.72, 80.28 to 233.49 and 92.37 to 253.98 mg/dl in T_0, T_1, T_2 and T_3 respectively. The average blood cholesterol concentration was 112.83, 137.20, 147.68 and 164.02 in group T_0, T_1, T_2 and T_3 respectively. The cholesterol level was significantly higher (P<0.05) in group T_3 and T_2 as compared to T_0 and T_1. The cholesterol level was significantly lower in control group (T_0) over all the treatment groups.

Table 38: Cholesterol concentration (mg/dl) of cattle fed with or without protected fat and protein

Treat/Days	-30	-15	0	30	60	90	Mean
T_0	89.16	84.64	65.81	102.35	155.63	179.39	**112.83^{c}**
T_1	86.60	103.09	76.48	158.66	193.63	204.72	**137.20^{b}**
T_2	82.42	103.66	80.28	166.31	219.91	233.49	**147.68^{ab}**
T_3	92.37	111.56	92.60	180.85	252.78	253.98	**164.02^{a}**
Mean	**87.64^{cd}**	**100.74^{c}**	**78.79^{d}**	**152.04^{b}**	**205.49^{a}**	**217.90^{a}**	**140.43**
	S	T	SxT	CV%			
SEm	7.358	6.008	14.715				
CD5%	20.395	16.652	NS	20.093			

abc bearing different superscripts within row and coloum different significantly (P<0.5)

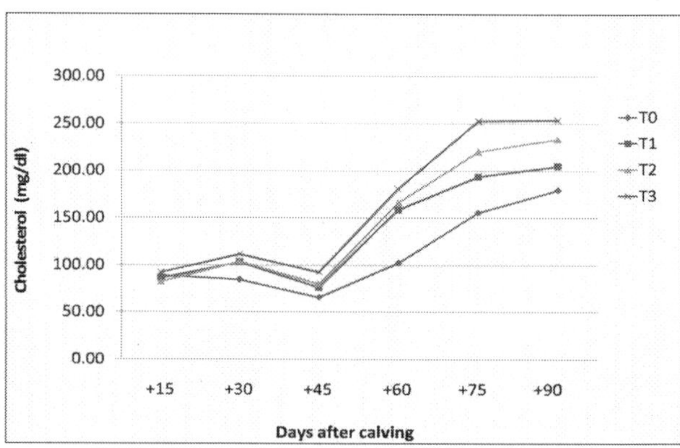

Fig. 25: Cholesterol concentration (mg/dl) of cattle fed with or without protected fat and protein

Higher cholesterol concentration is associated with better reproductive performance in high yielding dairy cows, as it acts as a precursor of steroid hormones (Son *et al.*, 1996). This higher level of cholesterol may have favorable effect on the synthesis of reproductive hormones like progesterone level (Wadhwa *et al.*, 2012) which might be responsible for better reproductive performance of protected fat group. Similar observation was reported by Staples *et al.* (1998) Zhang *et al.* (2011); Ranjan *et al.* (2012); Singh *et al.* (2014) and Kirovski *et al.* (2015) on supplementing bypass fat. Wheareas, Yadav *et al.* (2015) reported significantly decreased cholesterol level by feeding prilled fat to cows. Delbecchi *et al.* (2001) who observed non-significant increase in plasma cholesterol in multiparous, mid lactation Holstein cows supplemented with 4.8% formaldehyde protected canola seeds. However, by feeding bypass fat and protein significant improvement in cholesterol level was observed by Grewal *et al.* (2014) whereas, non significant difference was observed by Shelke *et al.* (2012) and Garg *et al.* (2012) by feeding bypass fat and protein.

4.11.5 Blood glucose

The blood glucose concentration (mg/dl) depicted in table 39 and varied from 57.28 to 71.41, 61.20 to 79.43, 60.73 to 74.36 and 62.74 to 89.12 mg/dl in treatment T_0, T_1, T_2 and T_3, respectively. The average blood glucose was 65.09, 67.76, 67.22 and 70.65 mg/dl in T_0, T_1, T_2 and T_3, respectively. The blood glucose level remains significantly higher in treatment groups over control whereas T_3, T_1 and T_2 were at par over control and T_0, T_1 and T_2 were at par over T_3. Similar results were reported by Zhang *et al.* (2011) and Singh *et al.* (2014) on supplementing bypass fat. Funstan *et al.* (1995) reported that fat supplementation might increase the glucose production through increased propionate production. This increse in glucose may have a possitive effect on LH relese. Garg *et al.* (1998) reviewed that UDP did not influence blood glucose level. Whereas, no difference was observed by Shelke *et al.* (2010) by feeding bypass fat and protein.

Table 39: Glucose concentration (mg/dl) of cattle fed with or without protected fat and protein

Treat/days	-30	-15	0	30	60	90	Mean
T_0	57.28	71.41	69.49	62.36	67.10	62.88	**65.09^b**
T_1	64.74	64.36	79.43	61.20	73.96	62.84	**67.76^{ab}**
T_2	60.73	74.36	69.80	67.16	68.11	63.18	**67.22^{ab}**
T_3	62.74	71.58	89.12	67.92	65.43	67.12	**70.65^a**
Mean	**61.37^d**	**70.42^b**	**76.96^a**	**64.66^{cd}**	**68.65^{bc}**	**64.00^{cd}**	**67.68**
	S	T	SxT	CV%			
SEm	2.025	1.653	4.049				
CD5%	5.612	4.63	NS	15.773			

abc bearing different superscripts within row and coloum different significantly (P<0.5)

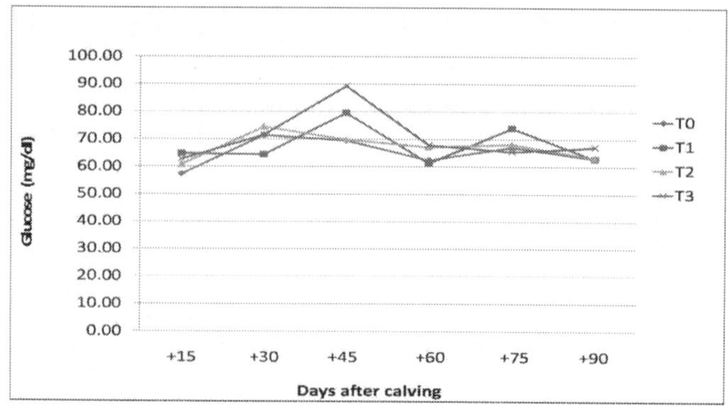

Fig. 26: Glucose concentration (mg/dl) of cattle fed with or without protected fat and protein

4.11.6 NEFA (Non Esterified Fatty Acid)

The NEFA (Table 40) varied from 0.97 to 1.54, 0.96 to 1.41, 0.83 to 1.22 and 0.86 to 1.39 mmol/l in T_0, T_1, T_2 and T_3, respectively during various periods. The overall mean of NEFA was 1.22, 1.24, 1.02 and 1.12 in T_0, T_1, T_2 and T_3, respectively which did not differ statistically in treatment groups. The results of the present study are in tune with the findings of Sklan *et al.* (1994); Fahey *et al.* (2002) and Tyagi *et al.* (2009a) who reported statistically non-significant increase in NEFA on supplementing Ca salts of FA. Tyagi *et al.* (2009) and Singh *et al.* (2014) reported increased and decreased trend respectively by feeding bypass fat. Sai *et al.* (2014) observed no difference in NAFA level by feeding protected amino acids. Whereas Delbecchi *et al.* (2001) reported significant increase in plasma NEFA levels on supplementing Formaldehyde protected canola seeds to lactating dairy cows. Also Shelke *et al.* (2010) reported no difference in NEFA level by feeding bypass fat and protein to buffalos.

Table 40: Non esterified fatty acid (NEFA) concentration (mmol/l) of cattle fed with or without protected fat and protein

Treat/days	-30	-15	0	30	60	90	Mean
T_0	1.54	1.29	1.40	1.09	1.02	0.97	**1.22**
T_1	1.31	1.08	1.41	1.39	1.28	0.96	**1.24**
T_2	1.22	0.99	0.87	1.06	1.16	0.83	**1.02**
T_3	1.39	1.06	1.08	0.97	0.86	1.37	**1.12**
Mean	**1.36^{a}**	**1.10^{ab}**	**1.19^{ab}**	**1.13^{ab}**	**1.08^{ab}**	**1.03^{b}**	**1.15**
	S	T	SxT	CV%			
SEm	0.110	0.090	0.220				
CD5%	0.303	NS	NS	55.788			

abc bearing different superscripts within row and coloum different significantly (P<0.5)

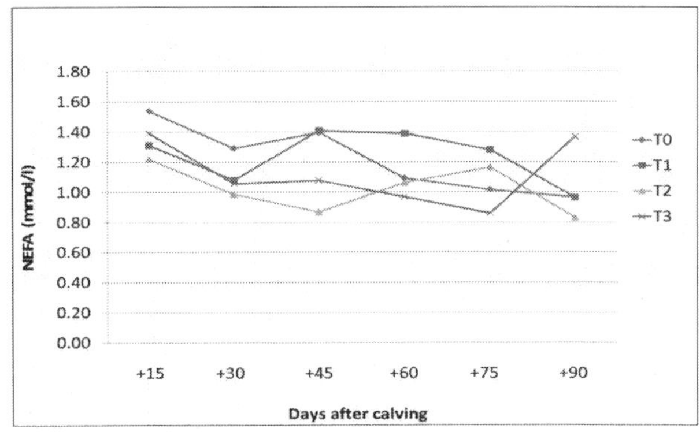

Fig. 27: Non esterified fatty acid (NEFA) concentration (mmol/l) of cattle fed with or without protected fat and protein

4.11.7 Total blood protein

Total blood protein concentration (mg/dl) is depticted in table 41 and ranged from 6.07 to 8.31, 6.23 to 8.64, 6.86 to 8.06 and 6.62 to 7.97 mg/dl in T_0, T_1, T_2 and T_3, respectively in different periods. The average mean of total blood protein was 7.35, 7.46, 7.21 and 7.21 mg/dl in T_0, T_1, T_2 and T_3, respectively. Highest total blood protein was observed in T_1 followed by T_0, T_3 and T_2. However, there was no significant difference between the groups. The period affected significantly in average total protein level on the animals. Similar non significant observations were reported by Garg (1998); Movaliya *et al.* (2013) by feeding bypass protein, Shelke *et al.* (2010) and Grewal *et al.* (2014) by feeding bypass fat and protein. Whereas, Sai *et al.* (2014) reported significantly increase in total protein level by feeding bypass protein.

Table 41: Total blood protein concentration (mg/dl) of cattle fed with or without protected fat and protein

Treat/days	-30	-15	0	30	60	90	Mean
T_0	6.07	6.81	6.60	8.31	8.27	8.04	**7.35**
T_1	6.37	6.84	6.23	8.64	8.51	8.19	**7.46**
T_2	6.91	6.86	6.86	8.06	7.52	7.04	**7.21**
T_3	7.03	7.14	6.62	7.52	7.97	7.01	**7.21**
Mean	**6.59^b**	**6.91^b**	**6.57^b**	**8.13^a**	**8.07^a**	**7.57^a**	**7.31**
	S	T	SxT	CV%			
SEm	0.203	0.166	0.407				
CD5%	0.564	NS	NS	12.400			

abc bearing different superscripts within row and coloum different significantly (P<0.5)

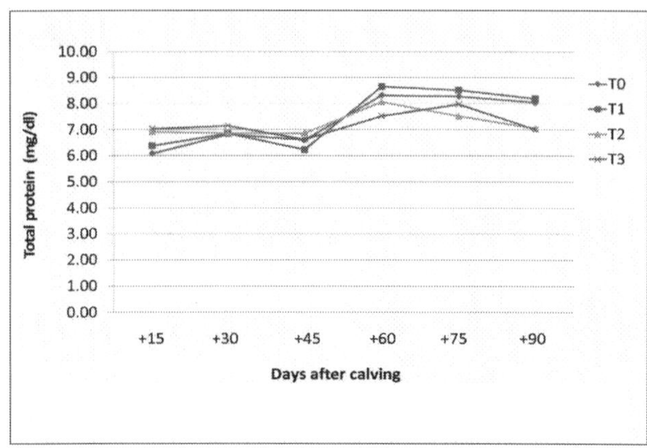

Fig. 28: Total blood protein concentration (mg/dl) of cattle fed with or without protected fat and protein

4.11.8 Triglyceride

Serum triglyceride concentration (mg/dl) is presented in table 42 and ranged between 6.79 to 18.30 in T_0, 8.52 to 16.94 in T_1, 9.79 to 20.39 in T_2 and 12.81 to 21.87 in T_3. The highest (P<0.05) mean value (17.01) was observed in T_3 followed by T_2 (15.26), T_1 (12.71) and T_0 (12.65) mg/dl, respectively.

Table 42: Triglyceride concentration (mg/dl) of cattle fed with or without protected fat and protein

Treat/days	-30	-15	0	30	60	90	Mean
T_0	11.19	9.68	6.79	13.47	16.49	18.30	**12.65^b**
T_1	13.67	11.15	8.52	11.46	14.52	16.94	**12.71^b**
T_2	14.94	14.72	9.79	14.30	17.44	20.39	**15.26^a**
T_3	12.81	14.10	12.94	19.49	21.87	20.83	**17.01^a**
Mean	**13.15^b**	**12.42^b**	**9.51^c**	**14.68^b**	**17.58^a**	**19.12^a**	**14.41**
	S	T	SxT	CV%			
SEm	0.929	0.759	1.858				
CD5%	2.576	2.103	NS	24.876			

abc bearing different superscripts within row and coloum different significantly (P<0.5)

The triglyceride level was significantly on higher side in bypass fat supplementing group T_2 and T_3 which was due to enhanced uptake of dietary fatty acids (Sklan and Tinsky 1993).

Barley and Baghel (2009) and Zhang *et al.* (2011) were observed similar results by supplementing bypass fat whereas, Tyagi *et al.* (2009a); Singh *et al.* (2014) and Kirovski *et al.* (2015) observed non significant difference by feeding bypass fat. Sai *et al.* (2014) reported that triglyceride level was significantly increased by feeding bypass protein. Also Shelke *et al.* (2010) and Grewal *et al.* (2014) observed non significant triglyceride by feeding bypass fat and protein.

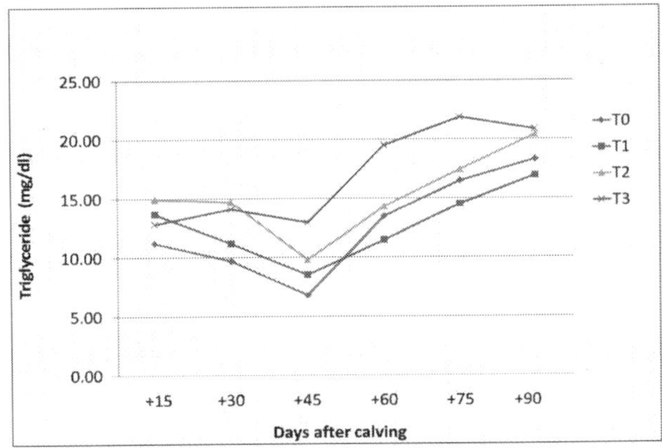

Fig. 29: Triglyceride concentration (mg/dl) of cattle fed with or without protected fat and protein

4.11.9 Uric acid

The blood uric acid value (Table 43) varied between 1.57 to 2.93 in T_0, 1.72 to 3.17 in T_1, 1.50 to 3.75 in T_2 and 1.97 to 3.05 in T_3. The higher mean value observed in T_3 (2.35) followed by in T_1 (2.32), T_2 (2.31) and T_0 (2.29) mg/dl, respectively which was non significant difference in all groups. Wadhwa *et al.* (2012) reported the blood uric acid levels increased ($P<0.05$) in the animals fed bypass fat.

Table 43: Uric acid concentration (mg/dl) of cattle fed with or without protected fat and protein

Treat/days	-30	-15	0	30	60	90	Mean
T_0	2.10	1.61	1.57	2.93	2.58	2.93	**2.29**
T_1	2.18	1.72	1.90	3.07	1.90	3.17	**2.32**
T_2	1.76	1.73	1.50	2.53	2.57	3.75	**2.31**
T_3	1.97	2.15	1.85	2.34	2.73	3.05	**2.35**
Mean	**2.00^{cd}**	**1.80^{d}**	**1.70^{d}**	**2.72^{ab}**	**2.45^{bc}**	**3.23^{a}**	**2.32**
	S	T	SxT	CV%			
SEm	0.201	0.164	0.401				
CD5%	0.556	NS	NS	33.526			

abc bearing different superscripts within row and coloum different significantly ($P<0.5$)

Whereas, Shelke *et al.* (2010) reported no difference in treatment by feeding bypass fat and protein.

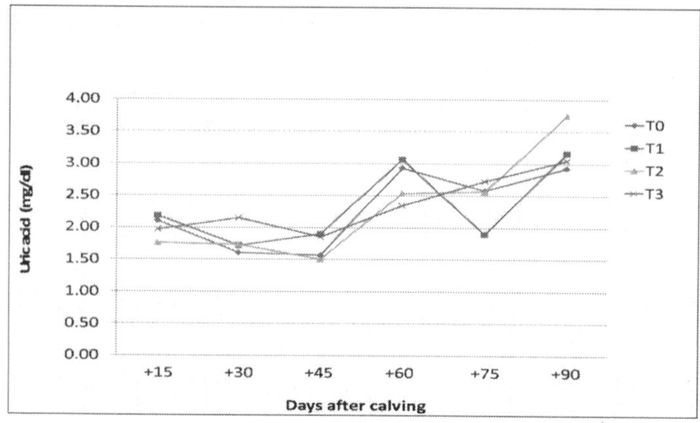

Fig. 30: Uric acid concentration (mg/dl) of cattle fed with or without protected fat and protein

4.11.10 Creatinine

The creatinine value (Table 44) ranged from 0.77 to 1.43 in T_0, 0.68 to 1.17 in T_1, 0.0.63 to 1.45 in T_2 and 0.65 to 1.35 mg/dl in T_3 groups. The average creatinine value was 0.94, 0.92, 0.98 and 0.97 mg/dl in T_0, T_1, T_2 and $T_{3,}$ respectively with non significant difference among the groups. These results were in agreement with the findings of Wadhwa *et al.* (2012) who reported that bypass fat supplementation had no effect on the creatinine.

Table 44: Creatinine concentration (mg/dl) of cattle fed with or without protected fat and protein

Treat/days	-30	-15	0	30	60	90	Mean
T_0	1.43	1.00	0.85	0.78	0.80	0.77	**0.94**
T_1	1.17	1.03	1.05	0.77	0.80	0.68	**0.92**
T_2	1.45	1.30	0.97	0.77	0.78	0.63	**0.98**
T_3	1.35	1.05	1.07	0.88	0.80	0.65	**0.97**
Mean	**1.35**^{a}	**1.10**^{b}	**0.98**^{c}	**0.80**^{d}	**0.80**^{d}	**0.68**^{c}	**0.95**
	S	T	SxT	CV%			
SEm	0.040	0.032	0.079				
CD5%	0.110	NS	NS	25.277			

abc bearing different superscripts within row and coloum different significantly (P<0.5)

Similarly Shelke *et al.* (2010) and Grewade *et al.* (2014) also reported no difference by feeding bypass fat and protein.

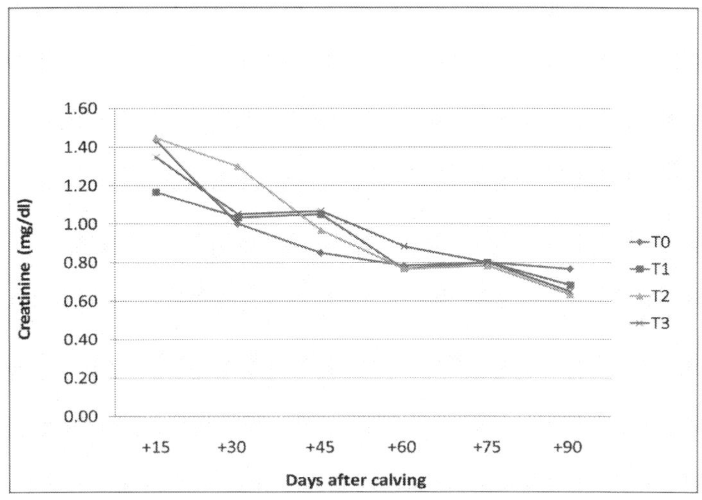

Fig. 31: Creatinine concentration (mg/dl) of cattle fed with or without protected fat and protein

4.12 Economics of Feeding Protected Fat and Protein in Crossbred Cows

The cost of formaldehyde treatment of groundnut cake @ 1.0 g/100g CP presented in Table 45. The economics of feeding protected fat and protein has been shown in Table 46. Net return over feed cost of milk yield per animal per day in T_0, T_1, T_2 and T_3 group was Rs. 183.51, 217.03, 196.04 and 208.00, respectively. The net return over feed was higher in T_1 (Rs. 33.52) follow T_3 (24.49) and T_2 (Rs. 12.53) over T_0. Similarly, net return over feed cost of 4% FCM yield per animal per day in T_0, T_1, T_2 and T_3 was Rs. 173.22, Rs. 209.48, Rs. 199.28 and Rs. 235.29, respectively. There was higher return in T_3 (Rs. 62.07) followed by T_1 (Rs. 36.26), T_2 (26.06) over T_0.

Table 45: Cost of formaldehyde treatment for groundnut cake

Cost of 40% formalin	Rs. 158/ ltr
Level of formaldehyde was @ 1.0 g FA/100 CP = 25 ml formalin/kg CP	2.5ml FA /100g CP
CP content of GNC	38% (0n DM basis)
100g of GNC	38 g of CP
100 kg of GNC	38 kg of C
1 qt GNC (38 kg CP) needs 38 * 25	0.950 liter formalin
Cost of 0.95 liter formalin = 0.950 * Rs. 158	Rs. 150.10
Labour cost for applying FA to 1qt GNC	Rs. 20

Total cost of FA treatment to 1 qt GNC	Rs. 170.10/kg
Concentration mixture includes 20% GNC, then additional cost for using FA treated GNC instead of untreated GNC is Rs. 34/ qtl concentrate mixture or 34 paise/ kg concentrate mixture	
Cost of rumen protected fat	Rs. 125

Feeding of the indigenously prepared bypass fat to dairy animals has shown to give additional profit of Rs. 34.50/- per cow per day (Naik *et al.*, 2009b), Rs. 11.60/- per cow per day (Gowda *et al.*, 2013) and Rs. 39.66/- per buffalo per day (Parnerkar *et al.*, 2011); Rs. 50/- per cow/ day, (Singh *et al.*, 2014) 15.97% (Sontakke *et al.*, 2014) and Rs. 94.46/ cow/ day, (Yadav *et al.* 2015) besides improvement in reproductive performance and health of the animals.

Table 46: Economics of feeding protected fat and protein to cattle during experimental period (120 days)

Attributes	T_0	T_1	T_2	T_3
Feed intake				
Green Maize intake (kg/d/animal)	12.375	12.300	11.891	12.266
Total green maize intake in 120 days trial by 6 animals (kg)	1485	1476	1426.92	1471.92
Green Lucern intake (kg/d/animal)	9.98	10.00	9.724	9.974
Total green lucern intake in 120 days trial by 6 animals (kg)	1197.6	1200	1166.88	1196.88
Green sugarcane intake (kg/d/animal)	4.379	4.412	4.11	4.641
Total green sugarcane intake in 120 days trial by 6 animals (kg)	525.48	529.44	493.2	556.92
Dry jowar intake (kg/d/animal)	3.521	3.516	3.323	3.521
Total dry jowar intake in 120 days trial by 6 animals (kg)	422.52	421.92	398.76	422.52
Concentrate intake (kg/d/animal)	5	5	5	5
Total concentrate intake in 120 days trial by 6 animals (kg)	600	600	600	600
Feed cost (Rs.)				
Total cost for green maize @ Rs.150/qtl	2227.5	2214.0	2140.38	2207.88
Total cost for green lucern @ Rs.195/qtl	2335.32	2340.00	2275.416	2333.916
Total cost for sugarcane @ Rs.200/qtl	1050.96	1058.88	986.40	1113.84
Total cost for dry jowar @ Rs.350/qtl	1478.82	1476.72	1395.66	1478.82
Total cost for concentrate @ Rs.2374/qtl,	14244	14448	14244	
FA treated Rs.2408	14448			
Total feed cost (Rs)	**21336.6**	**21537.6**	**21041.86**	**21582.46**
Cost of bypass fat supplementation @ 10 g/Lit Milk/animal @Rs 125/kg			82.08	89.28
Total cost of bypass fat for 120 days			10260	11160
Total feed cost with bypass fat	**21336.6**	**21537.6**	**31301.86**	**32742.46**
Average milk yield (ltr/d/animal)	**9.82**	**11.76**	**11.41**	**12.43**
Total milk yield in 90 days by 6 animals (liter)	5481	6350	6159	6518
Gross income (Rs.) from selling of milk @ Rs. 28/lit milk	**153468**	**177800**	**172452**	**182504**

Attributes	T_0	T_1	T_2	T_3
Average 4% FCM milk yield (ltr/d/animal)	**9.66**	**11.40**	**11.56**	**13.37**
Total 4% FCM milk yield in 90 days by 6 animals (liter)	5216.4	6156.0	6242.4	7219.8
Gross income (Rs.) from selling of milk @ Rs. 28/lit milk 4% FCM	**146059**	**172368**	**174787**	**202154**
Net return over feed cost (Gross income- Total feed cost)				
Milk yield				
Net return in 90 days from 6 animals (Rs.)	132131.4	156262.4	141150.1	149761.5
Net return in 90 days (Rs./ animals)	22021.9	26043.73	23525.02	24960.26
Net return (Rs./animal/day)	**183.51**	**217.03**	**196.04**	**208.00**
% increased over T_0	-	**15.44**	**6.39**	**11.77**
4% FCM Yield				
Net return in 90 days from 6 animals (Rs.)	124722.6	150830.4	143485.3	169411.9
Net return in 90 days (Rs./ animals)	20787.1	25138.4	23914.22	28235.32
Net return (Rs./ animal/day)	**173.22**	**209.48**	**199.28**	**235.29**
% increased over T_0	-	**17.31**	**13.08**	**26.38**

Also feeding bypass protein to dairy animals has shown to give additional profit of Rs. 9.61/- per day per animal (Gulati *et al.* (2002), Rs. 14.49/- per day per cow, Garg *et al.* (2003), Rs. 15.80 to 17.81/- per cow/ day (Sampath *et al.*, 2005), Rs. 12.75/- per animal per day (Mishra *et al.* (2006), Rs. 24.85/- per animal per day (Torane *et al.* 2006), Rs. 12.40/- per animal per day (Chandrashekharan *et al.*, 2008) and Rs. 23.70/- per animal per day (Amrutkar *et al.* (2014). Garg and Sherasia (2010) reviewed that feeding bypass protein meals, daily net income per animal increased by Rs. 10-11 in animals yielding 8-10 liters and Rs. 5-6 in animals yielding 4-5 liters. Gerg *et al.* (2002), Vahora *et al.* (2013) and Gajera *et al.* (2013) reported additional profit by feeding bypass fat and protein. Shelke *et al.* (2011b) and Grewal *et al.* (2014) recorded additional Rs. 46.91 and Rs. 11.05/- per animal per day profit by feeding bypass fat and protein.

The preferece indices for different parameters of study as affected by treatments were prepared by giving priority index and rank and are presented in table no 47. The best performance in response to the particular treatment was given a score of 100 and relative to it the scores for other treatments for respective parameters were calculated. The average of the relative scores obtained by each treatment for all studied parameters was thus obtained for deciding the best dietary treatment having overall better performance. The overall preference score was 81.26, 89.30, 93.52 and 97.64 % in T_0, T_1, T_2 and T_3, respectively. Highest score was observed in T_3 followed by T_2, T_1 and T_0.

Table 47: Preference indices of different feeding treatments (%)

Treatments	Body	Digestibility	Milk	Milk	Reproduction	Blood	Economics	Overall
		weight		yield	Composition	parameters		
T_0	71.30 (4)	109.80 (4)	75.57 (4)	86.66 (4)	48.06 (4)	89.26 (4)	79.88 (4)	**81.26 (4)**
T_1	84.85 (3)	112.46 (3)	89.19 (2)	91.34 (3)	68.24 (3)	93.38 (3)	94.48 (2)	**89.30 (3)**
T_2	87.14 (2)	115.40 (2)	88.92 (3)	91.74 (2)	96.53 (1)	94.04 (2)	87.26 (3)	**93.52 (2)**
T_3	100.00 (1)	117.57 (1)	100.00 (1)	98.60 (1)	90.58 (2)	96.50 (1)	97.81 (1)	**97.64 (1)**

* Figures in parentheses are rank of the treatment

Chapter 5

Summary and Conclusions

Twenty four crossbred cows in second to fourth lactation with most probable production ability (MPPA) of average approximately 2300 liter milk production per lactation for each group were selected from the herd maintained at RCDP on Cattle. The experimental animals were randomly divided into four treatment groups of 6 in each. Control group (T_0) was fed with 2/3 DM through roughages (2/3 from dry roughages and 1/3 from green roughages) + 1/3 DM from concentrate mixture (Conc. Mix includes 20% GNC (untreated). T_1 group was feed with 2/3 DM through roughages (2/3 from dry roughages and 1/3 from green roughages) + 1/3 DM from concentrate mixture (Conc. Mix includes 20% GNC treated with formaldehyde (FA) @ 1.0 g FA/100g CP). T_2 group was feed with 2/3 DM through roughages (2/3 DM from dry roughages and 1/3 DM from green roughages) + 1/3 DM from concentrate mixture (Conc. Mix includes 20% GNC untreated) + Bypass Fat(99%) @ 10g/lit milk production) and T_3 group was feed with 2/3 DM through roughages (2/3 DM from dry roughages and 1/3 DM from green roughages) + 1/3 DM from concentrate mixture (Conc. Mix includes 20% GNC treated with formaldehyde (FA) @ 1.0 g FA/100g CP) + Bypass Fat (99%) @ 10g/lit milk production). The duration of experiment was 120 days from 30 days prepartum to 90 days postpartum. The data generated during experimental period was analyzed by Factorial Randomized Block Design (FRBD)/ SAS, 9.3 versions. The results obtained during the course of this study have been summarized.

5.1 Proximate Chemical Analysis

5.1.1 Feed and fodder analysis

The values of percent DM, CP, CF, EE, TA and NFE for maize green fodder were 17.76, 09.85, 31.90, 1.68, 8.94 and 47.63 per cent, Lucerne 23.73, 21.54, 25.29, 1.85, 11.26 and 40.06 per cent, sugar cane were 33.93, 5.24, 31.66, 0.68,

8.18 and 54.24 and dry jawar were 60.68, 4.94, 33.74, 0.96, 11.84 and 48.52 per cent, respectively.

5.1.2 Experimental ration

The concentrate mixture of control group T_0 contained 88.15, 18.61, 13.41, 2.57, 7.85 and 57.56 per cent DM, CP, CF, EE, TA and NFE, respectively. The corresponding values for T_1, T_2 and T_3 were 88.10, 18.65, 13.45, 2.55, 7.85 and 57.50, 88.00, 18.40, 13.65, 2.62, 7.86 and 57.47 and 88.20, 18.55, 13.71, 2.61, 7.86 and 57.27 per cent, respectively.

5.1.3 Fatty acid profile of protected fat

The fatty acid profile of experimental protected fat was contains 85.66 per cent saturated fat, 13.28 per cent mono unsaturated fat and 1.03 per cent poly unsaturated fat.

5.2 Body Weight Changes During Experimental Period

Initial body weight of experimental animal was 438.33, 441.67, 405.00 and 458.33 kg in T_0, T_1, T_2 and T_3, respectively which were non significantly differed in all treatment. There was higher net loss of 13.3 kg in T_0 and higher weight gain was seen in T_3 (22 kg) followed by T_2 (15) and T_1 (4.17 kg), respectively after 6th fortnight period of calving. Gain in groups T_3 and T_2 attained positive energy balance from second fortnight onwards as was evident from increasing trend in body weight while it took longer (3rd fortnight) in case of group T_1 and (4th fortnight) in group T_0. Feeding protected fat and protein shows effective in reducing the extent and duration of body weight loss as compared to control group.

5.3 Body Condition Score During Experimental Period

Higher BCS observed in T_3 (3.64) followed by T_2 (3.60), T_1 (3.57) and T_0 (3.48), respectively which was statistically significant ($P<0.05$) over T_0 in all treatment.

5.4 Calving Performance

Average calf birth weight of calves born of experimental cows was higher in T_3 (25.5 kg) followed by T_1 (23.83 Kg), T_0 (23.17 Kg) and T_2 (20.67kg) which was statistically non-significant among treatment.

5.5 Digestibility Coefficient of Experimental Feed

The digestibility coefficients of DM were 58.49, 56.51, 59.43 and 61.04 per cent in T_0, T_1, T_2 and $T_{3,}$ respectively while that of crude protein were 62.04, 64.50, 63.12 and 65.20 per cent in T_0, T_1, T_2, and $T_{3,}$ respectively. In both parameters the differences were non-significant between groups. The digestibility coefficients of crude fiber were 62.43, 66.55, 66.18 and 67.27 per cent, respectively. There was statistically significant different ($P<0.05$) over control (T_0) whereas T_1, T_2 and T_3 were at par which indicating that feeding of protected protein or fat or combination of both promotes the fiber digestion. The values of higher digestibility coefficient of EE were significantly ($P<0.05$) in T_2 (68.41), T_3 (68.24), T_1 (66.59) and T_0 (62.07), respectively. The groups T_2, T_3, T_1 were at par over control T_0. The digestibility coefficients of NFE were 58.52, 57.17, 62.19 and 63.46 per cent in T_0, T_1, T_2 and $T_{3,}$ respectively. NFE digestibility was highest in treatment of feeding protected fat and protein combination (T_3) and followed by feeding of protected fat alone (T_2). Treatment (T_3) and (T_2) were at par from each other and treatment T_1 had significantly lower NFE digestibility.

5.6 Nutrient Intake

The DMI was 12.56, 12.59, 12.23 and 12.72 kg/day in T_0, T_1, T_2 and $T_{3,}$ respectively which was significant ($P<0.05$) higher in T_3 (12.72) followed by T_1 (12.59), T0 (12.47) and T_2 (12.23). However the DMI/ 100 kg body weight was 3.05, 3.01, 3.13 and 2.81 kg/d in T_0, T_1, T_2 and $T_{3,}$ respectively did not differ among all groups. The average digestible crude protein intake (DCPI) was 1.15, 1.21, 1.18 and 1.23 kg/d in T_0, T_1, T_2 and $T_{3,}$ respectively which was significantly higher ($P<0.01$) in T_3, T_2 and T_1 than T_0 due to higher DMI. However DCPI/ 100 kg body weight were 0.28, 0.29, 0.30 and 0.27 kg/d in T_0, T_1, T_2 and $T_{3,}$ respectively. There was no difference among all groups. Average total digestible nutrients intake (TDN) were 7.11, 7.86, 7.93 and 8.32 kg/day in T_0, T_1, T_2 and $T_{3,}$ respectively which was significantly higher ($P<0.01$) in T_3 followed by T_2 and T_1 over T_0 may be due to supplementation of protected fat and protein. The TDNI/ 100 kg body weight were 1.74, 1.88, 2.03 and 1.84 kg/day in T_0, T_1, T_2 and T_3 treatment, respectively showing the similar trend as that of DCP and DCPI/ 100kg body weight.

5.7 Milk Yield

5.7.1 Average milk yield

The average milk production during supplementation period was 9.82, 11.76, 11.41 and 12.43 kg/d in group T_0, T_1, T_2 and $T_{3,}$ respectively which was significantly 16.49% higher in T_1, 13.93% higher in T_2 and 20.99% in T_3 over $T_{0.}$

5.7.2 4% Fat corrected milk yield (4% FCM)

The average daily 4% FCM yield was 9.66, 11.40, 11.56 and 13.37 kg/d in T_0, T_1, T_2 and T_3 respectively. Group T_1, T_2 and T_3 had 15.26, 16.43 and 27.74% higher FCM yield over T_0 group. Significantly higher FCM was observed in T_3 over all treatment groups and T_1 and T_2 were at par whereas T_0 was significantly lower overall treatment group.

5.7.3 Energy corrected milk yield (ECM)

The daily energy corrected milk yield during supplementation period was 9.54, 11.08, 11.20 and 12.68 kg/d in group T_0, T_1, T_2 and T_3 respectively which was significantly higher 24.76% in T_3, 14.82% higher in T_2 and 13.89% in T_1 over T_0 Whereas treatment T_2 and T_1 were at par and significantly lower ECM found in T_0.

5.8 Milk Composition

5.8.1 Milk fat

The overall average milk fat percent was significantly higher (P<0.05) in group T_3 (4.52) than T_2 (4.03), T_0 (3.84) and T_1 (3.81). However, T_0, T1, and T_2 were at par to each other. Total fat yield kg/ day was significantly higher in T_3 (0.56) than T_2 (0.46), T_1 (0.44) and T_0 (0.39) which were at par.

5.8.2 Milk protein

The average protein content in T_0, T_1, T_2, and T_3 were 3.06, 2.96, 2.97 and 3.03 percent, respectively. The protein content observed significantly higher in T_0 followed by T_3 and almost same in T_2 and T_1 treatment group. Whereas, total protein yield kg/day significantly higher in T_3 (0.38) followed by T_1 (0.35), T_2 (0.34) than T_0 (0.30).

5.8.3 Milk lactose

The mean values were 4.93, 4.76, 4.77 and 4.80 in T_0, T_1, T_2, and T_3 respectively. The lactose content was significantly higher in T_0 over T_1, T_2 and T_3, respectively. Whereas the total lactose yield was significantly higher in T_3 (0.59) followed by T_1 (0.56) and T_2 (0.55) than control T_0 (0.48).

5.8.4 Total solids

The mean values of total solids were 12.60, 12.33, 12.56 and 13.16 in T_0, T_1, T_2 and T_3 respectively which was significantly higher in T3 treatment groups over T_0, T_1, and T_2 treatment group. Whereas, total yield of total solid kg/day was significantly higher in T_3 (1.63) than T_1 (1.45) and T_2 (1.44) but at par with each other and significantly lower in T_0 (1.24).

5.8.5 Solids not fat

The average values of SNF were 8.76, 8.52, 8.53 and 8.65 in T_0, T_1, T_2, and $T_{3,}$ respectively which was significantly higher in T_0 and T_3 over T_1 and T_2. Whereas total yield of SNF kg/day was significantly higher in T_3 (1.07) followed T_1 (1.00) and T_2 (0.98) over T_0 (0.86).

5.8.6 Milk density

The average density of milk of T_0, T_1, T_2, and T_3 were 29.88, 28.95, 29.10 and 28.63, respectively which did not differ from period (15) to period (90) days. However, it differs significantly between different treatments. Highest density (29.88) of milk was observed in To (control) group. Whereas at par in T1 and T_2 (28.95 and 29.10) and lowest was in T_3 (28.63).

5.8.7 Milk temperature

The average milk temperature in T_0, T_1, T_2, and T_3 were 25.79, 24.54, 24.95 and 24.35 ^{0}c, respectively which differ significantly higher T_0 over treatment groups.

5.8.8 Freezing point

The average freezing point in T_0, T_1, T_2, and T_3 were 0.58, 0.56, 0.56 and 0.56 ^{0}c, respectively which did not differ between the all groups.

5.8.9 Mineral matter

The average protein content in T_0, T_1, T_2, and T_3 were 0.79, 0.77, 0.77 and 0.76 percent, respectively which did not differ between the all groups.

5.8.10 Fatty acid profile

The total saturated fatty acid content in T_0, T_1, T_2 and T_3 were 68.32, 67.73, 65.88 and 65.24 percent, respectively. The total unsaturated fatty acid content in T_0, T_1, T_2 and T_3 were 26.85, 28.03, 29.00 and 29.58 percent, respectively. The monounsaturated fatty acid in T_0, T_1, T_2 and T_3, were 21.24, 22.18, 22.75 and 23.28 percent, respectively. The poly unsaturated fatty acid in T_0, T_1, T_2 and T_3 were 5.61, 5.85, 6.25 and 6.30 per cent, respectively. Overall, there was increased concentration of mono, poly and unsaturated fatty acid by supplementing bypass fat and protein.

5.9 Reproduction Performance

5.9.1 Expulsion of fetal membrane (EFM)

The average time required for expulsion of fetal membrane (EFM) was 5.81, 4.11, 3.51 and 3.93 hrs in group T_0, T_1, T_2 and T_3 respectively. There was significantly less ($P<0.05$) time was required for EFM in T_2 (2.3), T_3 (1.88) and T_1 (1.7) over than T_0.

5.9.2 Involution of uterus

The involution of uterus was completed in 28.33, 27.50, 25.17 and 22.50 days in T_0, T_1, T_2 and T_3 respectively. There were ($P<0.05$) significantly less days required for involution of uterus in T_3 (5.83 days) and T_2 (3.16 days) than T_0 and T_1 (0.83).

5.9.3 Open period

The average duration for commencement of heat was 87.83, 76.00, 66.00 and 61.83 days in T_0, T_1, T_2 and T_3 respectively. The time required for commencement of first heat after calving was reduced by 26 days in T_3, 21.83 days in T_2 and 11.83 days in T_1 over T_0. There was statistically significant ($P<0.01$) difference in T_3 and T_2 over T_0 and T_1 and T_0 were at par.

5.9.4 Service period

The average service period were significantly lower in T_2 (82.00) than all treatment groups followed by T_3 (99.67), T_1 (130.67) and T_0 (141.33) days.

5.9.5 Number of AI

The average numbers of AI required were 3.00, 2.50, 1.83 and 2.00 in T_0, T_1, T_2 and T_3, respectively. Significantly less number of AI were required in all treatment groups over control.

5.9.6 Conception rate

Highest conception rate per cent was found in T_2 (66.67) treatment group followed by T_3 (54.55), T_1 (40.00) and lowest in T_0 (33.34).

5.10 Blood Parameters

The overall average was 3.28, 3.21, 3.21 and 3.24 mg/dl in respective groups. The albumin level remained within normal range and no difference was observed at any stage within treatment of experiment groups. The overall mean was

4.06, 4.26, 3.99 and 3.97 in T_0, T_1, T_2 and $T_{3,}$ respectively. There was no statistically difference in all treatments. The average values were 16.57, 13.69, 15.98 and 14.48 mg/dl in T_0, T_1, T_2 and $T_{3,}$ respectively. The average BUN level was lower significantly ($P<0.01$) in T_1 and T_3 compare to T_0 and T_2 treatments groups. The treatment and period effect on BUN concentration of the animals. The average was 112.83, 137.20, 147.68 and 164.02 in group T0, T_1, T_2 and $T_{3,}$ respectively. The cholesterol level was higher in group T_3 and T_2 as compare to T_0 and T_1 which was statistically significant. The cholesterol level was significantly lower in control group (T_0) over all treatment groups. The average blood glucose was 65.09, 67.76, 67.22 and 70.65 mg/dl in T_0, T_1, T_2 and T_3, respectively. The blood glucose level remain higher significantly in treatment groups over control whereas T_3, T_1 and T_2 were at par over control and T_0, T_1 and T_2 were at par over T_3. The overall mean of NEFA was 1.22, 1.24, 1.02 and 1.12 mmol/l in T_0, T_1, T_2 and $T_{3,}$ respectively which did not differ in treatment groups. The average mean of total protein was 7.35, 7.46, 7.21 and 7.21 mg/dl in T_0, T_1, T_2 and $T_{3,}$ respectively. However, there was no significant difference between the groups. The period affected on average total protein level on the animals. The highest ($P<0.05$) mean value of triglycerides was observed in T_3 (17.01) followed by T_2 (15.26), T_1 (12.71) and T_0 (12.65) mg/dl, respectively. The higher mean value of NEFA was observed in T_3 (2.35) followed by in T_2 (2.31), T_1 (2.32) and T_0 (2.29) mg/dl, respectively which was no significant difference in all groups. The higher mean value of uric acid observed in T_3 (2.35) followed by in T_1 (2.32), T_2 (2.31) and T_0 (2.29) mg/dl, respectively which was non significant difference in all groups. The average creatinine value was 0.94, 0.92, 0.98 and 0.97 mg/dl in T_0, T_1, T_2 and $T_{3,}$ respectively which showed no difference in groups.

5.11 Economic

Net return over feed cost of milk yield per animal per day in T_0, T_1, T_2 and T_3 group was Rs. 183.51, 217.03, 196.04 and 208.00, respectively. The net return over feed was higher in T_1 (Rs. 33.52) follow T_3 (24.49) and T_2 (Rs. 12.53) over T_0. Similarly, net return over feed cost of 4% FCM yield per animal per day in T_0, T_1, T_2 and T_3 was Rs. 173.22, Rs. 209.48, Rs. 199.28 and Rs. 235.29, respectively. There was higher return in T_3 (Rs. 62.07) followed by T_1 (Rs. 36.26), T_2 (26.06) over T_0.

Conclusions

Based on the results of the present investigation the following conclusions are drawn

- Feeding of either protected fat or protected protein improved digestibility of nutrients. However, when protected fat and protein were fed in combination, the digestibility of carbohydrates was increased significantly.
- Milk production was significantly enhanced by feeding either protected protein or protected fat alone or in combination with much alteration in milk composition.
- Feeding of protected fat separately or in combination with protected protein helped cows to recover BCS faster in the post-partum phase and also helped cows to start reproduction cycle earlier. Hence, it is concluded that feeding protected fat helped to increase overall reproductive ability of animals.
- By supplementation of protected fat, blood cholesterol and triglycerides level were significantly increased where as by supplementing protected protein BUN level significantly decreased.

Therefore, feeding of protected fat or protected protein and combination of both improves digestibility of nutrients and milk production, respectively. Feeding protected fat helped to increase overall reproductive ability. However, if fed in combination, it helps to recover body condition score earlier. Hence, it could be concluded that, feeding of combination of protected protein and protected fat in early lactation is more remunerative from production and reproduction point of view.

Literature Cited

Ambasankar, K. and Balakrishnan, V. 2011. Influence of protected sardine oil on in vitro rumen fermentation and nutrient digestibility of complete diet. Indian J. Anim. Sci., 81(1): 84-86.

Amrutkar, S.A., Thakur, S.S. and Pawar, S.P. 2014. Economics of supplementing rumen protected methionine and lysine in the ration of lactating crossbred cows. Indian J. Anim. Nutr., 31 (1) : 14-19.

AOAC, 2005. Association of Official Analytical Chemists. Official Methods of Analysis. Washington, DC.

Arewad, G. R., Thube, H. A., Pandya, P. R., Parnerkar S. and Shankhpal S. 2011. Effect of feeding bypass protein based total mixed ration on performance of growing crossbred calves. Indian J. Anim. Nutr. 2011.28 (3) : 303-308.

Bahram R., Safdar, A. H. A. and Kor, N. M. 2014. Mechanisms through which fat supplementation could enhance reproduction in farm animal. Euro. J. Exp. Bio., 4(1):340-348.

Barley, G.G. and Baghel, R.P.S. 2009. Effect of bypass fat supplementation on milk yield, fat percentage and serum triglyceride levels of Murrah buffaloes. Buffalo Bulletin 28 (4): 173-175.

*Bhosale, A.B., Gadegaonkar, G.M., Patil, M.B., Kank, V.D., Hol, B.G. and Gulavane, S.U. 2010. Effect of bypass fat supplementation on the fat corrected milk yield and feed efficiency in lactating buffaloes. Proc. of International Buffalo Conference, Vol: 11, 1-4 Feb 2010, New Delhi. P: 111. FAO Electronic conference on successes and failures with animal nutrition practices and technologies in developing countries. 11: 61-64.

Campanile, G., De Filippo, C., Di Palo, R., Taccone, W. and Zicarelli, L. 1998. Influence of dietary protein on urea levels in blood and milk of buffalo cows. Livestock Prod. Sic., 55: 135-143.

Cant, J.P., DePeters, E.J. and Baldwin, R.L. 1991. Effect of dietary fat and postruminal casein administration on milk composition of lactating dairy cows. J. Dairy Sci., 74:211.

Chalupa, W., Vecchiarelli B., Elser, A.E., Kronfeld, D.S., Sklan, D. and Palmquist, D.L. 1986. Ruminal fermentation in vivo as influenced by long chain fatty acids. J. Dairy Sci., 69: 1293-1301.

Chandrasekharaiah, M., Sampath, K.T. and Praveen, U.S. 2008. Effect of bypass protein on milk production performance in crossbred cows. Indian J. Anim. Sci., 78(5): 527-530.

Chatterjee, A. and Walli, T.K. 2003. Effect of formaldehyde treatment on effective protein degradability and *in vitro* post-ruminal digestibility of mustard cake. Indian J. Anim. Nutr., 20(2): 143-148.

Chatterjee, A. and Walli, T.K. 1998. Effect of feeding formaldehyde treated mustard cake on growth performance of Murrah buffalo calves. Indian J. Anim. Prod. Mgmt. 39:163.

Chouinard, P.Y., Girard, V. and Brisson, G.J. 1997. Lactational response of cows to different concentrations of calcium salts of canola oil fatty acids with or without biocarbonates. J. Dairy Sci., 80: 1185-1193.

Chouinard, P.Y., Girard, V. and Brisson, G.J. 1998. Fatty acid profile and physical properties of milk fat from cows fed calcium salts of fatty acids with varying unsaturation. J. Dairy Sci., 81: 471-481.

CSO. 2011. Central Statistical Organisation, Govrtment of India.

Dairy Technical Service Staff. 2002. Enertia PFA calcium salts of palm fatty acids (PFA), Rumen Bypass Fat. The Official Answer Guide. ADM Animal Health and Nutrition, 1000 N. 30th Quincy, IL 62301, 877-236-2460.

Delbecchi, L., Ahnadi, C.E. Kennelly, J.J. and Lacasse, P. 2001. Milk fatty acid composition and mammary lipid metabolism in Holstein cows fed protected or unprotected conola seeds. J. Dairy Sic., 84: 1375-1381.

DePeters, E.J. and Cant, J.P. 1992. Nutritional factors influencing the nitrogen composition of bovine milk: A review. J. Dairy Science, 75: 2043-2070.

Dhali, A., Mehal, R.K., Sirohi, S.K., Mech, A. and Karunakaran, M. 2006. Monitoring feeding adequacy in dairy cows using milk urea and milk protein contents under farm condition. Asian-Aust. J.Anim Sic. 19(12): 1742-1748.

Dhulipalla, S. K., Venkateswarlu S. and Raja K. K. 2013. Effect of supplementation of Milk Dhara on milk yield and milk composition in graded Murrah buffaloes. J. Adv. Vet. Res. 3: 89-92.

Dosky, K. N. S., Jaaf, S. S. A. and Mohammed, L. T. 2012. Effect of protected soybean meal on milk yield and composition in local Meriz goats. Mesopotamia J. of Agric. 40(1):1-10.

Ekeren, P.A., Smith, D.R., Lunt, D.K. and Smith, S.B. 1992. Ruminal biohydrogenation of fatty acids from high oleate sunflower seeds. J. Anim. Sci., 70: 2574-2580.

Elliott, J. P., Overton, T. R. and Drackley, J. K.. 1994. Digestibility and effects of three forms of mostly saturated fatty acids. J. Dairy Sci. 77:789.

Fahey, J., Mee J.F., Murphy, J.J. and Callaghan, D.O, 2002. Effects of calcium salts of fatty acids and calcium salt of methionine hydroxyl analogue on plasma prostaglandin F2á metabolite and milk fatty acid profile in late lactation Holstein Friesian cows. Theriogenology, 58: 1471-1482.

FAO STAT. 2012. Food and Agriculture Organization, Statistical Data 2012.

Ferguson, J.D., Galligan, D.T. and Thomsen, N. 1994. Principal descriptors of body condition score in Holstein cows. J. Dairy Sci., 77: 2965.

Ferguson, J.D., Sklan, D., Chalupa, W. and Kronfied, D.S. 1990. Effect of hard fats on in vitro and In vivo rumen fermentation, milk production and reproduction in dairy cows. J. Dairy Sci., 73: 2864-2879.

Fiorentini G., Santana M. C. A., Sampaio A. A. M., Reis R. A., Ribeiro A. F., Berchielli T. T. 2012. Intake and performance of confined crossbred heifers fed different lipid sources. Revista Brasileira de Zootecnia., 41(6): 1490-1498.

Funston, R.N., Roberts, A.J., Hixon, D.L., Hallford, D.M., Sanson, D.W. and Moss G.E. 1955. Effect of acute glucose antagonism on hypophyseal hormones and concentrations of insulinlike growth factor (IGF)-1 and IGF binding proteins in serum, anterior pituitary and hypothalamus of ewes. Biol. Reprod., 52:1179-1186.

Gajera, A.P., Dutta, K.S., Parsana, D.K., Savsani, H.H., Odedra, M.D., Gajbhiye, P.U., Murthy, K.S. and Chavda, J.A. 2013. Effect of bypass lysine, methionine and fat on growth and nutritional efficiency in growing Jaffrabadi heifers. Vet. World 6(10): 766-769.

Garcia-Bojalil, C.M., Staples, C.R., Risco, C.A., Savio, J.D. and Thatcher, W.W. 1998. Protein degradability and calcium salts of long chain fatty acids in the diets of lactating dairy cows: productive responses. J. Dairy Sci., 81: 1374-1384.

Garg, M. R., Sherasia, P. L., Bhanderi, B.M., Gulati, S. K. and Scott, T.W. 2007. Milk production efficiency improvement in buffaloes through the use of slow ammonia release and protected protein supplement. Italian J. Anim. Sci., 6(2):1043-1045.

Garg, M.R. 1998. Role of bypass protein in feeding ruminants on crop residue based diet– review. Asian-Australasian J. Anim. Sci., 11: 107-116.

Garg, M.R. and Mehta, A.K. 1998. Effect of feeding bypass fat on feed intake, milk production and body condition of Holstein Friesian cows. Indian J. Anim. Nutri., 15(4): 242-245.

Garg, M.R. and Sherasia, P.L. 2010. Rumen bypass protein technology for enhancing productivity in dairy animals. Anim. Production and Health, FAO Electronic conference, 1-30 September, 2010. 11: 61-64.

Garg, M.R., Bhanderi, B.M. and Sherasia, P.L. 2012. Effect of supplementing bypass fat with rumen protected choline chloride on milk yield, milk composition and metabolic profile in crossbred cows. Indian J. Dairy Sci. 65(4): 319-323.

Garg, M.R., Sherasia, P.L. and Bhanderi, B.M. 2012. Effect of supplementing bypass fat with and without rumen protected choline chloride on milk yield and serum lipid profile in Jaffarabadi buffaloes. Buffalo Bulletin. 31(2): 91-98.

Garg, M.R., Sherasia, P.L., Bhanderi, B.M., Gulati, S.K., Ashes, J.R. and Scott, T.W. 2004. Effect of feeding rumen protected protein on milk production and composition of lactating cows. Indian Vet. J. 81: 48-50.

Garg, M.R., Sherasia, P.L., Bhanderi, B.M., Gulati, S.K. and Scott, T.W. 2002. Efffect of feeding rumen protected protein on milk production in lactating cows. Indian J. Anim. Nutri., 19(3): 191-198.

Garg, M.R., Sherasia, P.L., Bhanderi, B.M., Gulati, S.K. and Scott, T.W. 2003[a]. Effect of feeding rumen protected protein on milk production in lactating buffaloes. Anim. Nutri. Feed Tech., 3: 151-157.

Garg, M.R., Sherasia, P.L., Bhanderi, B.M., Gulati, S.K. and Scott, T.W. 2003[b]. Effect of feeding rumen protected protein on milk production in lactating cows. Indian J. Dairy Sci., 56: 218-222.

Garg, M.R., Sherasia, P.L., Bhanderi, B.M., Gulati, S.K., Scott, T.W. 2005. Effect of Feeding Rumen Protected Protein on Milk Production in Low Yielding Crossbred Cows. Anim. Nutri. and Feed Tech., 5(1): 1-8.

Goff, J.P. and Horst, R.L. 1997. Physiological changes at parturition and their relationship to metabolic disorders. J. Dairy Sci. 80:1260-1268.

Gowda, N.K.S., Manegar, A., Raghavendra, A., Verma, S., Maya, G., Pal, D.T., Suresh, K.P. and Sampath, K.T. 2013. Effect of protected fat supplementation to high yielding dairy cows in field condition. Anim. Nutri. and Feed Tech., 13(1): 125-130.

Grewal, R.S., Tyagi, N., Lamba, J.S., Ahuja, C.S. and Saijpaul, S. 2014. Effect of bypass fat and niacin supplementation on the productive performance and blood profile of lactating crossbred cows under field conditions. Anim. Nutri. and Feed Tech., 14: 573-581.

Grummer, R.R. 1988. Influence of prilled fat and calcium salt of palm oil fatty acids on ruminal fermentation and nutrient digestibility. J. Dairy Sci., 71: 117-123.

Grummer, R.R. 1995. Impact of Changes in organic nutrient metabolism on feeding the transition dairy cows. J. Anim. Sci., 73:2820-2833.

Gulati, S.K., Garg, M.R., Serashia, P.L., Scott, T.W., 2003. Enhancing milk quality and yield in the dairy cow and buffalo by feeding protected nutrient supplements. Asian Pacific J. Clinical Nutri. 12: 61-63.

Gulati, S.K., Scott, T.W., Garg, M.R. and Singh, D.K. 2002. An overview of rumen protected proteins and their potential to increase milk production in India. Indian Dairyman, 54(3): 31-35.

Gutierrez Castaneda E., Overton, T.R., Butler, W.R. and Bauman, D.E. 2005. Dietary supplements of two doses of calcium salts of conjugated linoleic acid during the transition period and early lactation. J. Dairy Sci., 88: 1078-1089.

ICAR (1998). Nutrient requirements of livestock and poultry, 2nd revised edition. Indian Council of agricultural Research, New Delhi.

Jadhav, S. E., Pattanaik, A. K., Dutta, N., Garg, A.K. and Verma, A.K. 2014. Nutrition, productivity and product quality in farm animals. Center of Advanced Faculty Training in Animal Nutrition, Indian Veterinary Research Institute, Izatnagar, India. 218 pp.

Jenkins, T.C. 1998. Fatty acid composition of milk from Holstein cows fed oleamide or canola oil. J. Dairy Sci., 81: 794-800.

Jenkins, T.C. 1999. Lactation performance and fatty acid composition of milk from Holstein cows fed 0 to 5% oleamide. J. Dairy Sci., 82: 1525-1531.

Jenkins, T.C. and Palmquist, D.L. 1982. Effect of added fat and calcium on in vitro formation of insoluble fatty acid soaps and cell wall digestibility. J. Anim. Sci., 55: 957-963.

Jenkins, T.C. and Palmquist, D.L. 1984. Effects of fatty acids or calcium soaps on rumen and total nutrient digestibility of dairy rations. J. Dairy Sci., 67: 978-986.

Kalabande, V.H. and Thomas, C.T. 1999. Effect of bypass protein on yield and composition of milk in crossbred cows. Indian J. Anim. Sci., 69(8): 614-616.

Karcagi, R.G., Gaál, T., Ribiczey, P., Huszenicza, G. and Husveth, F. 2010. Milk production, peripartal liver triglyceride concentration and plasma metabolites of dairy cows fed diets supplemented with calcium soaps or hydrogenated triglycerides of palm oil. J Dairy Res.,77(2):151-158.

Kaur, H. and Arora, S.P. 1995. Dietary effects on ruminant livestock reproduction with particular reference to protein. Nutri. Res. Reviews., 8:121-136.

Kirovski, D., Blond, B., Katiæ, M., Markoviæ, R. and Šefer, D. 2015. Milk yield and composition, body condition, rumen characteristics, and blood metabolites of dairy cows fed diet supplemented with palm oil. Chemical and Biological Technologies in Agriculture, 2 (1); PP: 1-5.

Kumar, S., Kumari, R., Kumar, K. and Walli, T K. 2015. Roasting and formaldehyde method to make bypass protein for ruminants and its importance: A review. Indian J. Anim Sci. 85 (3): 223–230.

Kumar, R. M., Tiwari, D.P. and Kumar, A. 2005. Effect of undegradable protein level and plane of nutrition on lactation performance on crossbred cattle, Asian-Australasian J. Anim. Sci., 18: 1407-1413.

Kunju, P.K.G., Mehta, A.K. and Garg, M.R. 1992. Feeding of bypass protein to crossbred cows in India on straw based ration. Asian-Australasian J. Anim. Sci., 5(1): 107-112.

Lounglawan, P., Suksombat, W. and Chullanandana, K. 2007. The effect of ruminal bypass fat on milk yields and milk composition of lactating dairy cow. Suranaree J. Sci. Tech. 14(1): 109-119.

Lush, J.L.1945. Animal Breeding Plans (2nd ed.) Collegiate Press, Inc. Ames, Iowa. Madson, J.1982 The effect of formaldehyde treated protein and urea on milk yield and composition in dairy cows. Acta Agric. Scandinavia., 32(4) : 389-395.

Mathew, J. J. and Zachariah A. S. 2014. Evaluating effects of rumen escape fat (REF) supplementation on early lactation in crossbred cows. Intas Polivet. 15 (I): 47-51.

McNamara, S., Butler, T., Ryan, D.P., Mee,J.F., Dillon, P. O., Mara, F.P., Butler, S.T., Anglessy, D., Rath, M. and Murphy, J.J. 2003 Effect of offering rumen protected fat supplements on fertility and performance in spring-calving Holstein-Friesian cows. Anim. Reprod, Sci., 79:45-46.

Mishra, B.B., Swain, R.K., Sahu, B. K. and Samantaray. 2006. Effect of bypass protein supplementation on nutrient utilization, milk production and its composition in crossbred cows on paddy straw based ration. Animal Nutri. and Feed Tech. 6(1): 123-33.

Mishra, S., Thakur, S.S., Raikwar, R. 2004. Milk production and composition in crossbred cows fed calcium salts of mustard oil fatty acids. Indian J. Anim. Nutri., 21(1): 22-25.

Morgan, D.J. 1985. The effect of formalin treated soybean meal upon the performance of lactating cows. Anim. Prod., 41 (1): 33-42.

Movaliya, J.K., Padodara, R.J., Bhadaniya, A.R. and Savsani, H.H. 2013. Effect of bypass methionine-lysine supplementation of haematologocal and biochemical parameters of Jaffrabadi heifers. Vet. World. 6(3): 147-150.

Mudgal, V., Baghel, R.P.S., Ganie, A. and Srivastava, S. 2013. Effect of feeding bypass fat on intake and production performance of lactating crossbred cows. Indian J. Anim. Res., 46(1): 103-104.

Naik, P.K. 2012. Feeding rumen protected fat to high yielding dairy cows. In: Animal Nutrition: Advances and Developments (Eds. U.R. Mehra, Putan Singh and A.K. Verma). Satish Serial Publishing House, Delhi, India, pp. 529-548.

Naik, P.K. 2013a. Bypass fat in dairy ration-A review. Anim. Nutri. and Feed Tech., 13(1): 147163.

Naik, P.K. 2013b. Benefits and scope of indigenously prepared bypass fat. Indian Dairyman, February Issue, pp. 66-68.

Naik, P.K. and Singh, N.P. 2011. Technology for preparation and feeding of bypass fat (rumen protected fat) to dairy animals. Extension Folder No. 46, ICAR Research Complex for Goa, Old Goa.

Naik, P.K., Dhuri, R.B., Swain, B.K., Karunakaran, M., Chakurkar, E.B. and Singh, N.P. 2013a. Analysis of existing dairy farming in Goa. Indian J. Anim. Sci., 83: 299-303.

Naik, P.K., Saijpaul, S. and Kaur, K. 2010. Effect of supplementation of indigenously prepared rumen protected fat on rumen fermentation in buffaloes. Indian J. Anim. Sci., 24:212- 215.

Naik, P.K., Saijpaul, S. and Rani, N. 2007a. Evaluation of rumen protected fat prepared by fusion method. Anim. Nutri. and Feed Tech., 7: 95-101.

Naik, P.K., Saijpaul, S. and Rani, N. 2007b. Preparation of rumen protected fat and its effect on nutrient utilization in buffaloes. Indian J. Anim. Nutri., 24(4): 212-215.

Naik, P.K., Saijpaul, S. and Rani, N. 2009a. Effect of ruminally protected fat on in vitro fermentation and apparent nutrient digestibility in buffaloes (Bubalus bubalis). Anim. Feed Sci. and Tech., 153: 68-76.

Naik, P.K., Saijpaul, S., Sirohi, A.S. and Raquib, M. 2009b. Lactation response of cross bred dairy cows fed indigenously prepared rumen protected fat - A field trial. Indian J. Anim. Sci., 79: 1045-1049.

Naik, P.K., Swain, B.K., Karunakaran, M. and Singh, N.P. 2013b. Production of bypass fat indigenously for dairy animals. ICAR News, 19:17.

Nam, I.S., Choi J.H., Seo K.M. and Ahn J.H. 2014. In vitro and lactation responses in mid lactating dairy cows fed protected amino acids and fat. Asian Austral. J. Anim. Sci. 27(12): 1705-1711.

NDDB. 2014. National Dairy Development Board.

NebGuide. 2004. Supplemental fat for high producing dairy cows. Institute of Agriculture and Natural Resources, University of Nebraska-Lincoln Extension, USA.

NRC. 1989. Nutrient requirements of dairy cattle, 6th ed. National Research Council, National Academy of Sciences, Washington, D.C.

NRC. 2001. Nutrient Requirements for Dairy Cattle, 7th rev. ed. National Academy of Sciences, Washington, DC.

Palmquist, D.L. 1991. Influence of source and amount of dietary fat on digestibility in lactating cows. J. Dairy Sci., 74: 1354-1360.

Palmquist, D.L. and Jenkins, T.C. 1980. Fat in lactation rations: Review. J. Dairy Sci., 63: 1-14.

Palmquist, D.L. and Moser, E.A. 1981. Dietary fat effects on blood insulin, glucose utilization and milk protein of lactating cows. J. Dairy Sci., 64: 1664.

Parnerkar, S., Kumar, D., Shankhpal, S.S. and Thube, M. 2011. Effect of feeding bypass fat to lactating buffaloes during early lactation. In: Proceedings of 14th Biennial Conference of Animal Nutrition Society of India 'Livestock Productivity Enhancement with Available Feed Resources', Nov. 3-5, 2011, Pantnagar, India, pp. 111-112.

Patel, V. R., Gupta, R. S., Parnerkar, S., Jani, V. R. and Garg, D. D. 2012. Performance of Buffalo Heifers fed on Bypass Protein: An On-farm Appraisal. Anim. Nutri. and Feed Tech., 12(3): 395- 402.

Prasad, S. 1994. Body condition scoring and feeding management in relation to production performance of crossbred dairy cattle. Ph.D. Thesis submitted to NDRI Deemed University, Karnal.

Purushothaman, S., Kumar, A. and Tiwari D. P. 2008. Effect of feeding calcium salts of palm oil fatty acids on performance of lactating crossbred cows. Asian-Aust. J. Anim. Sci. 21(3):376-385.

Qaisi Mohmmad A. Al- and Titi Hosam H. 2014. Effect of rumen-protected methionine on production and composition of early lactating Shami goats milk and growth performance of their kids. Archiv Tierzucht 57(1), 1-11.

Qureshi, M.S. and Tawheed, A.A. 2012. International Conference on Applied Life Sciences (ICALS2012) Turkey, September 10-12, 2012.

Ramteke, P.V., Patel, D.C., Parnerkar, S., Shankhpal, S.S., Patel, G.R., Katole, S.B. and Pandey, A. 2014. Effect of feeding bypass fat prepartum and during early lactation on productive performance in buffaloes. Livestock Research International, July-September, 2(3): 63-67.

Ranjan, A., Sahoo, B., Singh, V. K., Srivastava, S., Singh, S. P. and Pattanaik, A. K. 2012. Effect of bypass fat supplementation on productive performance and blood biochemical profile in lactating Murrah (Bubalus bubalis) buffaloes. Trop. Anim. Health Prod. 44(7): 1615-1621.

Ranjhan, S.K. 1998 Nutrient requirement of livestock and poultry, Indian Council of Agricultural Research (ICAR) Publication, New Delhi, India.

Remppis, S., Steingass, H. Gruber, L. and Schenkel, H. 2011. Effects of energy intake on performance, mobilization and retention of body tissue, and metabolic parameters in dairy cows with special regard to effects of prepartum nutrition on lactation: a review. Asian-Aust. J. Anim. Sci. 24:540-572.

Rice, V.A., Andrew, F.N., Warnick, E.J., Legates, J.E., 1970. Breeding and improvement of farm animals. 6th ed. Tata McGraw Hill Publishing Co., Bombay, India.

Sahoo, B. and Walli, T. K. 2007. Effect of feeding rumen degradable protein and supplemental energy on growth performance and blood metabolites in growing kids. Anim. Nutr. Feed Tech.,7:53-61.

Sahoo, B. and Walli, T.K. 2005. Effect of feeding bypass protein as formaldehyde treated mustard cake along with energy supplement on blood metabolites and milk production in lactating goats. Indian J. Dairy Sci., 58 (3):184-190.

Sahoo, B., Walli, T.K. and Sharma, A.K. 2006. Effect of formaldehyde treated rape seed oil cake based diet supplemented with molasses on growth rate and histopathological changes in goats. Asia. Austra. J. Anim. Sci., 19(7): 997-1003.

Sai, S., Thakur, S.S., Kewalramani, N. and Chaurasia, M. 2014. Effect of supplementation of rumen protected methionine plus lysine on growth performance, nutrient utilization and blood metabolites in calves. Indian J. Anim. Nutri. 31 (1): 1-7.

Saijpaul, S., Naik, P.K. and Rani N. 2010. Effects of rumen protected fat on in vitro dry matter degradability of dairy rations. Indian J. Anim. Sci., 80(10): 993-997.

Sampath, K.T., Prasad, C.S., Ramachandra, K.S., Sudresan K. and Subbarao, A. 1997. Effect of feeding undegraded dietary protein on milk production of crossbred cows. Indian J. Anim. Sci., 67(1): 706-708.

Sampath, K.T., Chandrasekharaiah, M. and Praveen, U.S. 2005. Effect of bypass protein on milk production of crossbred cows- A field study. Indian J. Anim. Nutri. 22: 41-43.

Sarwar, M., Sohaib, A., Ajmal Khan, M. and Nisa, M. 2003. Effect of feeding saturated fat on milk production and composition in crossbred dairy cows. Asian-Aust. J. Anim. Sci. 16(2): 204-210.

SAS. 2011. SAS version 9.3.1. SAS Institute Inc., Cary, NC, USA.

Schauff, D.J. and Clark, J.H. 1989. Effects of prilled fatty acids and calcium salts of fatty acids on rumen fermentation, nutrient digestibilities, milk production and milk composition. J. Dairy Sci., 72: 917-927.

Schauff, D.J. and Clark, J.H. 1992. Effects of feeding diets containing calcium salts of long chain fatty acids to lactating dairy cows. J. Dairy Sci., 75: 2990-3002.

Schneider, P., Sklan, D., Chalupa, W. and Kronfeld, D. S. 1988. Feeding calcium salts of fatty acids to lactating cows. J. Dairy Sci., 71: 2143-2150.

Scott, T.A., Shaver, R.D., Zepeda, L., Yandell, B. and Smith, T.R. 1995. Effects of rumen inert fat on lactation, reproduction and health of high producing Holstein herds. J. Dairy Sci., 78: 2435-2451.

Sengar, S.S.and Mudgal, V.D. 1982. Effects of feeding treated and untreated proteins on the growth rate pattern and nutrients utilization in kids. Indian J. Anim. Sci., 52(7):517-523.

Sharma, D.D. 2004. All India Dairy Husbandry Officers' Workshop, Dairy Extension Division, National Dairy Research Institute, Karnal, India, pp. 28-33.

Shelke, S. K. 2010. Milk production and immune status and reproductive performance of Murrah buffaloes supplemented with rumen protected fat and protein. A thesis submitted to the National dairy research institute, Karnal.

Shelke, S. K., Thakur, S. S. and Shete, S. M. 2012. Protected nutrients technology and impact of feeding protected nutrients to dairy animals: A review. Inter. J. of Dairy Sci. 7(3): 51-62.

Shelke, S. K., Thakur, S. S. and Shete, S. M. 2012a. Productive and reproductive performance of Murrah buffaloes (Bubalus bubalis) supplemented with rumen protected fat and protein. Indian J. Anim. Nutri., 29(4): 317-323.

Shelke, S.K. and Thakur, S.S. 2011a. Effect on the quality of milk and milk products in murrah buffaloes (*Bubalus bubalis*) fed rumen protected fat and protein. Int. J. Dairy Sci., 6(2): 124-133.

Shelke, S.K. and Thakur, S.S. 2011b. Economics of feeding rumen protected fat and protein to lactating Murrah buffaloes. Indian J. Anim. Nutri., 28(3): 278-282.

Shelke, S.K., Thakur, S.S. and Amrutkar, S.A. 2011. Effect of pre partum supplementation of rumen protected fat and protein on the performance of Murrah buffaloes. Indian J. Anim. Sci., 81: 946-950.

Shelke, S.K., Thakur, S.S and Amrutkar, S.A. 2012b. Effect of feeding protected fat and proteins on milk production, composition and nutrient utilization in Murrah buffaloes (*Bubalus bubalis*). Anim. Feed Sci. Techn., 171: 98-107.

Sherasia, P.L., Garg, M.R., Bhanderi, B.M. 2012. Effect of feeding slow ammonia release and protected protein supplement in lactating cows and buffaloes. Indian J. Anim. Res., 46(2): 168-171.

Singh, M., Sehgal, J. P., Roy, A. K., Pandita, S and Rajesh, G 2014. Effect of prill fat supplementation on hormones, milk production and energy metabolites during mid lactation in crossbred cows. Vet. World. 7(6): 384-388.

Sirohi, S. K., Walli, T.K., Garg, M.R. and Kumar, B. 2013. Effect of formaldehyde treated mustard cake on nutrient utilization and milk production performance in crossbred cows fed wheat straw based diet. Indian J. Anim. Nutri. 30 (1): 5-11.

Sirohi, S.K., Wali, T.K. and Mohanta, R. 2010. Supplementation effect of bypass fat on production performance of lactating crossbred cow. Indian J. Anim. Sci., 80(8): 733-736.

Sjaunja, L.O., Baerre, L., Junkkarinen, L., Pendersen, J., Setala J. 1990. A Nordic proposal for an energy corrected milk (ECM) formula. In 27th session International committee of recording and productivity of milk animal, Paris, France, PP. 156-157.

Sklan, D. and Tinsky, M. 1993 Production and reproduction responses by dairy cows fed varying undegradable protein coated with rumen bypass fat. J. Dairy Sci., 76:216-223.

Sklan, D., Kaim, M., Moallam, U. and Folman, Y. 1994. Effect of dietary calcium soaps on milk yield, body weight, reproductive hormones, and fertility in first parity and older cows. J. Dairy Sci., 77: 1652-1660.

Sklan, D., Moallem, U. and Folman, Y. 1991. Effect of feeding calcium soaps of fatty acids on production and reproductive responses in high producing lactating cows. J. Dairy Sci., 74: 510-17.

Snedecor, G.W. and Cochran, W.G. 1989. Statistical Methods. 7th Edn. Lowa State University Press, Ames, IA.

Son, J., Grant, R.J. and Lavson, L.L. 1996. Effect of follow and escape protein on lactational and reproductive performance of dairy cows. J. Dairy Sci., 79:822-830.

Sontakke, U.B., Kaur, H., Tyagi, A., Kumar, M. and Saikh, A.H. 2014. Effect of feeding rice bran lyso-phospholipids and rumen protected fat on feed intake, nutrient utilization and milk yield in crossbred cows. Indian J. Anim. Sci., 84 (9): 998-1003.

Staples, C.R., Burke, J.M. and Thatcher, W.W. 1998. Influence of supplemental fats on reproductive tissues and performance of lactating cows. J. Dairy Sci., 81: 856-871.

Strusiñska, D., Minakowski, D., Pysera, B. and Kaliniewicz, J. 2006. Effects of fat-protein supplementation of diets for cows in early lactation on milk yield and composition. Czech J. Anim. Sci., 51 (5): 196–204.

Sukhija, P.S. and Palmquist, D.L. 1990. Dissociation of calcium soaps of long-chain fatty acids in rumen fluid. J. Dairy. Sci. 73: 1784-1787.

Suresh, K.P., Bhatta, R., Mondal, S. and Sampath, K.T. 2011. Effect of bypass protein on milk yield in Indian cattle–A meta-analysis. Anim. Nutri. and Feed Tech. 11: 19-26.

Sutton, J.D., Knight, R., Mc Allan, A.B. and Smith, R.H. 1983. Digestion and synthesis in the rumen of sheep given diets supplemented with free and protected oils. British J. Nutri., 49: 419-432.

Thakur, S.S. and Shelke, S.K. 2009. Growth performance and nutrient utilization of Murrah buffalo calves fed ration supplemented with bypass fat prepared from soybean acid oil. Indian J. Anim. Sci., 79: 1238-1241.

Thakur, S.S. and Shelke, S.K. 2010. Effect of supplementing bypass fat prepared from soybean acid oil on milk yield and nutrient utilization in Murrah buffaloes. Indian J. Anim. Sci., 80: 354-357.

Tiwari, D.P. and Yadava, I.S. 1994. Effect on growth, nutrient utilization and blood metabolites in buffalo calves fed rations containing formaldehyde treated mustard cake. Indian J. Anim. Sci., 64(6): 625-630.

Torane, D.J., Wankhede, S.M. and Kalbande, V.H. 2006. Effect of feeding different levels and sources of bypass protein with urea treated wheat straw on performance of crossbred (Holstein Friesian x Deoni) calves. Anim. Nutri. and Feed Tech., 6: 79-86.

Treacher, R.J., Reid, I.M. and Roberts, C.J. 1986. Effects of body condition at calving on the health and performance of dairy cows. Anim. Prod. 43:1-6.

Tøináctý, J., Køí•ová, L., Hadrovál, S., Hanuš, O., Janštová, B., Vorlová, L. and Draèková, M. 2006. Effect of rumen-protected protein supplemented with three amino acids on milk yield, composition and fatty acid proûle in dairy cows. J. Anim. and Feed Sci.,. 15: 3–15.

Tyagi, N., Thakur, S.S. and Shelke, S.K. 2010. Effect of pre partum bypass fat supplementation on productive and reproductive performance in crossbred cows. Trop. Anim. Health Prod., 42: 1749-1755.

Tyagi, N., Thakur, S.S. and Shelke, S.K. 2009[a]. Effect of feeding bypass fat supplement on milk yield, its composition and nutrient utilization in crossbred cows. Indian J. Anim. Nutri., 26(1): 1-8.

Tyagi, N., Thakur, S.S. and Shelke, S.K. 2009[b]. Effect of pre-partum bypass fat supplementation on the performance of crossbred cows. Indian J. Anim. Nutri., 26(4): 247-250.

Vahora, S. G., Kore, K. B., and Parnerkar, S. 2012. Feeding of formaldehyde-treated protein meals to lactating buffaloes; effect on milk yield and composition. Livestock research for rural development. 24: 2.

Vahora, S.G. and Parnerkar, S. 2013. Effect of feeding bypass nutrients on milk production and composition in buffaloes under field conditions. Indian J. Anim. Nutri., 30(1): 67-71.

Vahora, S.G., Parnerkar, S. and Kore, K.B. 2013. Productive Efficiency of Lactating Buffaloes Fed Bypass Fat under Field Conditions: Effect on Milk Yield, Milk Composition, Body Weight and Economics. Iranian J. Applied Anim. Sci. 3(1): 53-58.

Wadhwa, M., Grewal, R.S., Bakshi, M.P.S. and Brar, P.S. 2012. Effect of supplementing bypass fat on the performance of high yielding crossbred cows. Indian J. Anim. Sci., 82(2): 200-203.

Walli , T. K. 2008. Bypass protein technology- A success story in feeding of dairy animals for increasing milk production at a cheaper cost. Indian Dairyman. 60: 53-60.

Walli, T.K. 2005. Bypass protein technology and the impact of feeding bypass protein to dairy animals in tropics: A Review. Indian J. Anim. Sci. 75 (1): 135-142.

*Walli, T.K. and Sirohi, S.K. 2004. Evaluation of heat treated (roasted) soybean on lactation crossbred cows. Project of the Collaborative Project between National Dairy Research Institute, Karnal and American Soybean Association, New Delhi, Feb. 2004. PP.

Wankhede, S.M. and Kalbande, V.H. 2001. Effect of feeding bypass protein with urea treated grass on the performance of Red Kandhari calves. Asian Aust. J. Anim. Sci., 14(7): 970-973.

West, J.W. and Hill, G.M. 1990. Effect of protected fat product on productivity of lactating Holstein and Jersey cows. J. Dairy Sci., 73: 3200-3207.

Wu, Z. and Palmquist, D. L. 1991. Synthesis and biohydrogenation of fatty acids by ruminal microorganisms in vitro. J. Dairy Sci., 74: 3035-3046.

Yadav, C. M. and Chaudhary, J. L. 2004. Effect of feeding protected protein on nutrient utilization, milk yield and milk composition of lactating crossbred cows. Indian J. Dairy Sci., 57: 394-399.

Yadav, C.M. and Chaudhary, J.L. 2010. Effect of feeding formaldehyde treated groundnut cake on dry matter intake, digestibility of nutrients and body measurements in crossbred heifers. Anim. Nutri. and Feed Tech. 10: 107-113.

Yadav, G., Roy, A. and Singh, M. 2015. Effect of Prilled Fat Supplementation on Milk Production Performance of Crossbred Cows. Indian J. Anim. Nutri. 32 (2): 133-138.

Yadav, N.S., Wankhede, S.M., Kalbande, V.H. 2006. Effect of feeding bypass protein with urea treated sorghum straw on performance of crossbred (Jersey x Red Kandhari) calves. Anim. Nutri. and Feed Tech., 6(1): 87-94.

Zhang, H., Wang, Z., Liu, G., He, J., Su, C. 2011. Effect of dietary fat supplementation on milk components and blood parameters of early-lactating cows under heat stress. Slovak J. Anim. Sci., 44(2): 52-58.

Colour Plates

Chapter 3: Material and Methods

Plate 1: Preparation of Protected protein by Formaldehyde treatment

Plate 2: Protected Fat

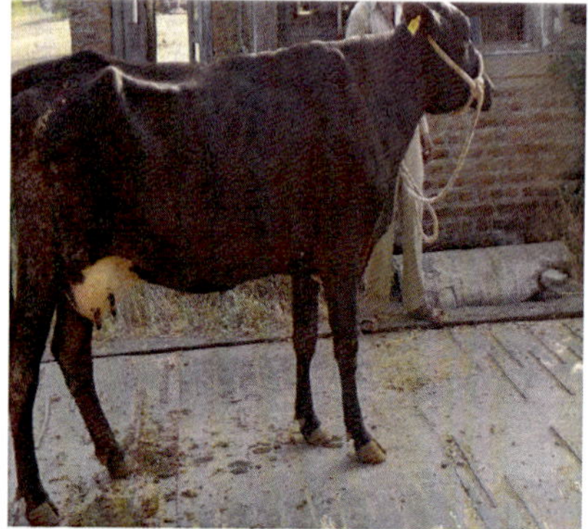

Plate 3: Weighing of experimental animal

Plate 4: Feeding of experimental animals

Plate 5: Blood serum separation

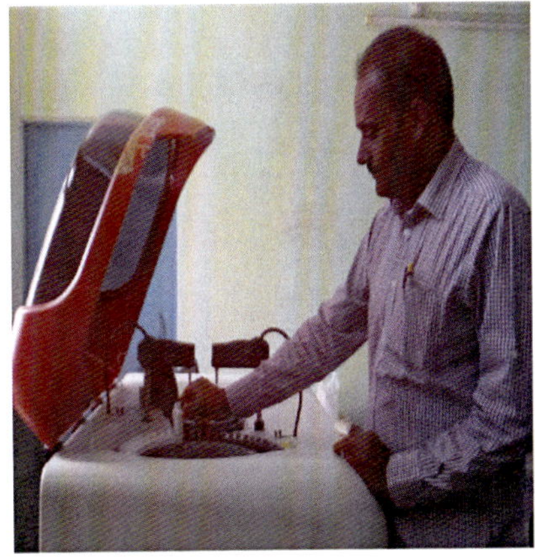

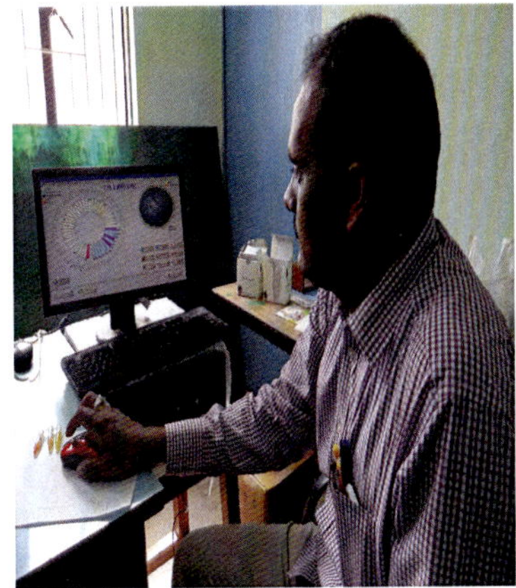

Plate 6: Estimation of blood profile of animal on automatic blood analyzer

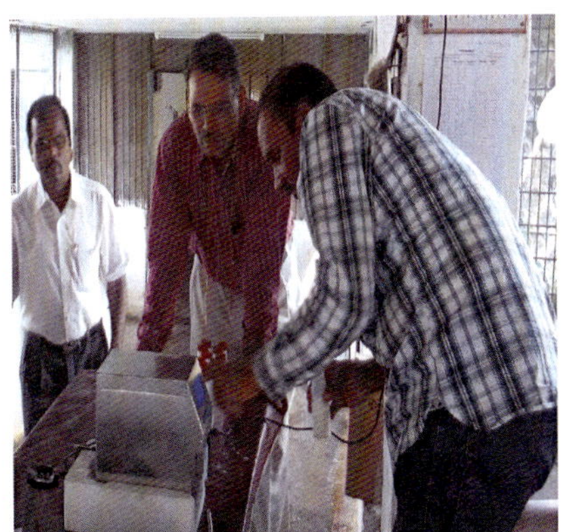

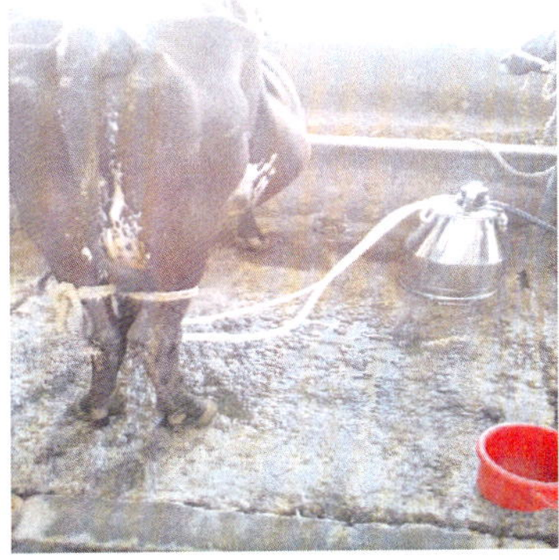

Plate 7: Estimation of milk composition on lactoscan

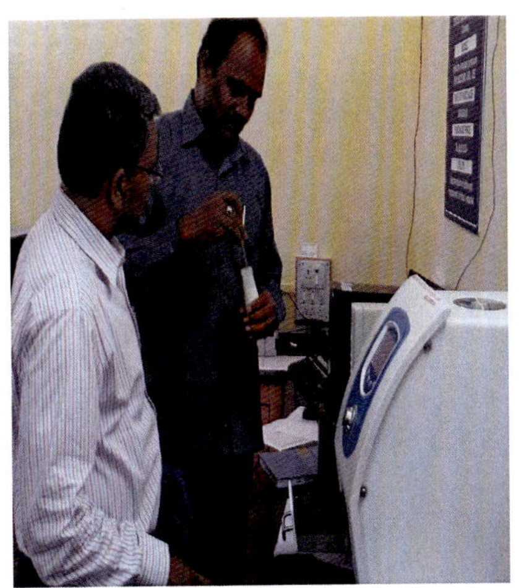

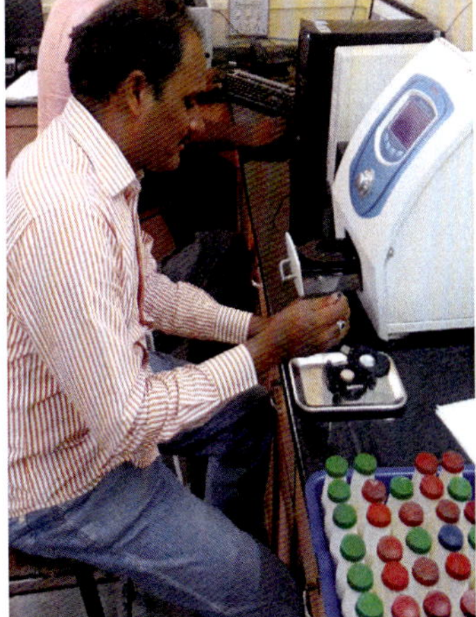

Plate 8: Estimation of fatty acid profile of milk on Spectra analyzer

Plate 9: Chemical analysis of feed and fodder ingredients

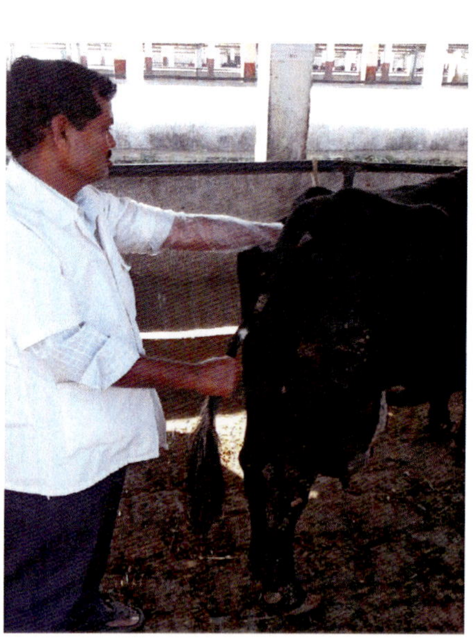

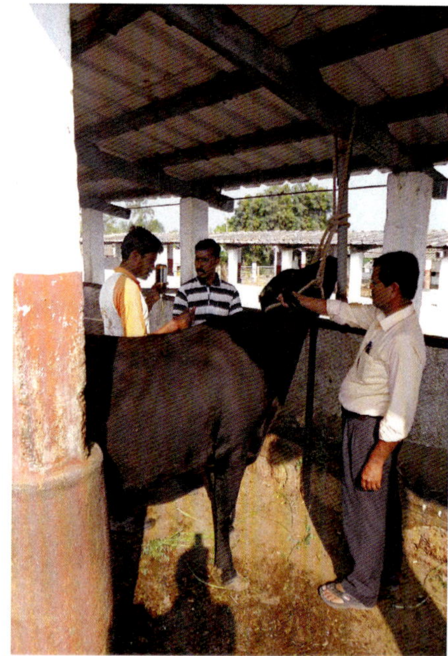

Plate 10: Study of reproduction performance and abnormalities

Appendix

S. No	Parameters	Results				Score (%)			
		T_0	T_1	T_2	T_3	T_0	T_1	T_2	T_3
A	**Body weight**								
1	weight gain after 90 days (kg)	-13.33	4.17	15	22	-60.59	18.95	68.18	100.00
2	BCS	3.48c	3.57b	3.60ab	3.64a	95.60	98.08	98.90	100.00
3	Calf weight (kg)	23.17	23.83	20.67	25.5	90.86	93.45	81.06	100.00
B	**Digestibility of Nutrients (%)**								
1	DM	58.49	56.51	59.43	61.04	95.82	92.58	97.36	100.00
2	CP	62.04	64.5	63.12	65.2	95.15	98.93	96.81	100.00
3	CF	62.43b	66.55a	66.18a	67.27a	92.81	98.93	98.38	100.00
4	EE	62.07b	66.59a	68.41a	68.24a	90.73	97.34	100.00	99.75
5	NFE	58.52bc	57.17c	62.19ab	63.46a	92.22	90.09	98.00	100.00
C	**Milk Production (kg)**								
1	Milk Yield	9.82b	11.76a	11.41a	12.43a	79.00	94.61	91.79	100.00
2	FCM	9.66c	11.40b	11.56b	13.37a	72.25	85.27	86.46	100.00
3	ECM	9.54c	11.08b	11.20b	12.68a	75.24	87.38	88.33	100.00
D	**Milk Composition**								
1	Milk Fat %	3.84b	3.81b	4.03b	4.52a	84.96	84.29	89.16	100.00
2	Fat Yield kg	0.39b	0.44b	0.46b	0.54a	72.22	81.48	85.19	100.00
3	Protein %	3.06a	2.96c	2.97bc	3.03ab	100.00	96.73	97.06	99.02
4	Protein Yield kg	0.30a	0.35b	0.34b	0.38b	78.95	92.11	89.47	100.00
5	Lactose %	4.93a	4.76b	4.77b	4.80b	100.00	96.55	96.75	97.36
6	Lactose Yield kg	0.48b	0.56a	0.55a	0.59a	81.36	94.92	93.22	100.00
7	Total Solid %	12.60b	12.33b	12.56b	13.16a	95.74	93.69	95.44	100.00
8	Solid yield kg	1.24c	1.45b	1.44b	1.63a	76.07	88.96	88.34	100.00
9	SNF %	8.76a	8.52b	8.53b	8.65ab	100.00	97.26	97.37	98.74
10	SNF yield kg	0.86b	1.00a	0.98a	1.07a	80.37	93.46	91.59	100.00

E	**Reproduction**								
1	EFM (hrs)	5.81a	4.11b	3.51b	3.93b	34.47	82.91	100.00	88.03
2	Involution of uterus (days)	28.33a	27.50a	25.17ab	22.50b	74.09	77.78	88.13	100.00
3	1st heat after calving (Days)	87.83a	76.00ab	66.00b	61.83b	57.95	77.08	93.26	100.00
4	Service period (Days)	141.33	130.67	82	99.67	27.65	40.65	100.00	78.45
5	AI/ conception	3.00a	2.50ab	1.83b	2.00b	36.07	63.39	100.00	90.71
6	Conception rate	33.34	40	66.67	54.55	50.01	60.00	100.00	81.82
F	**Blood constituents**								
1	Albumin	3.28	3.21	3.21	3.24	100.00	97.87	97.87	98.78
2	Globumin	4.06	4.26	3.99	3.97	95.31	100.00	93.66	93.19
3	BUN	16.57a	13.69b	15.98a	14.48b	78.96	100.00	83.27	94.23
4	Cholosterol	112.83c	137.20b	147.68ab	164.02a	68.79	83.65	90.04	100.00
5	Glucose	65.09b	67.76ab	67.22ab	70.65a	92.13	95.91	95.15	100.00
6	NEFA	1.22	1.24	1.02	1.12	80.39	78.43	100.00	90.20
7	Total Protein	7.35	7.46	7.21	7.21	98.53	100.00	96.65	96.65
8	Triglyciride	12.65b	12.71b	15.26a	17.01a	74.37	74.72	89.71	100.00
9	Uric acid	2.29	2.32	2.31	2.35	100.00	98.69	99.13	97.38
10	Creatinine	0.94	0.92	0.98	0.97	97.83	100.00	93.48	94.57
G	**Economics**								
1	Net return (Rs./animal/day)	183.516	217.0311	196.0419	208.002	84.88	100.00	90.03	95.77
2	Net return 4% FCM (Rs./animal/day)	173.225	209.486	199.285	235.29	74.55	88.60	84.30	100.00
	Total (out of 3900 marks)					**3014.73**	**3394.71**	**3623.55**	**794.65**
	100 scale (%)					**71.78**	**80.83**	**86.27**	**90.35**